AF576017

Physical and Mathematical Fluid Mechanics

Physical and Mathematical Fluid Mechanics

Editor

Markus Scholle

MDPI • Basel • Beijing • Wuhan • Barcelona • Belgrade • Manchester • Tokyo • Cluj • Tianjin

Editor
Markus Scholle
Heilbronn University
Germany

Editorial Office
MDPI
St. Alban-Anlage 66
4052 Basel, Switzerland

This is a reprint of articles from the Special Issue published online in the open access journal *Water* (ISSN 2073-4441) (available at: https://www.mdpi.com/journal/water/special_issues/physical_mathematical_fluid_mechanics).

For citation purposes, cite each article independently as indicated on the article page online and as indicated below:

LastName, A.A.; LastName, B.B.; LastName, C.C. Article Title. *Journal Name* **Year**, *Volume Number*, Page Range.

ISBN 978-3-03943-747-4 (Hbk)
ISBN 978-3-03943-748-1 (PDF)

Cover image courtesy of Markus Scholle.

Contents

About the Editor

Markus Scholle (Prof. Dr. rer. nat.) is working as a Professor of Physics at the Heilbronn University of Applied Sciences, Germany. He earned his Ph.D. in Physics (1999) from the University of Paderborn (title of the thesis: Das Hamiltonsche Prinzip in der Kontinuumstheorie nichtdissipativer und dissipativer Systeme—Ein neues Konzept zur Konstruktion von Lagrangedichten) at the Institute of Theoretical Physics and his postdoctoral lecture qualification in Mechanics (2004) from the University of Bayreuth (title of the professorial dissertation: Einfluß der Randgeometrie auf die Strömung in fluiden Schichten) in the Faculty of Engineering Science. Being both a physicist and engineer with a focus on fluid mechanics, his research activities at the Institute for Flow in Additively Manufactured Porous Media (ISAPS) are multidisciplinary at the interface between physics and engineering sciences. Particular research interests include modeling and simulation of coating and lubrication flows, potential methods in general fluid mechanics, variational calculus with discontinuities, and nonlinear acoustics. He has authored more than 80 publications, 39 of them being peer-reviewed.

Preface to "Physical and Mathematical Fluid Mechanics"

Fluid dynamics is one of the oldest physical disciplines. Looking at ancient times, the works of Archimedes and Sextus Iulius Frontinus can already be considered as relevant contributions due to their technological importance for shipbuilding and water supply. In the Renaissance, Leonardo da Vinci dealt with fluid dynamics, and relevant contributions were later made by Galileo Galilei, Evangelista Torricelli, Blaise Pascal, Edme Mariotte, Isaac Newton, and Daniel Bernoulli. In view of the close relationship between physics and mathematics, the works of Leonhard Euler and Jean-Baptiste le Rond d'Alembert can be regarded as pioneering for the subsequent development of fluid dynamics because the potential theory that emerged from them, which, in addition to the theory of frictionless and vortex-free flows, also enabled the calculation of electromagnetic fields, is a prime example of how physics and mathematics inspire each other. Since that time, fluid dynamics and physics have been inseparably linked. Despite the success of potential theory, its limitations were soon recognized, as became evident from various paradoxes such as the famous d'Alembert paradox. The consequent extension of Euler's theory with regard to viscosity by Claude Louis Marie Henri Navier and George Gabriel Stokes to the Navier–Stokes equations posed new challenges for mathematics, becoming subject to generations of scientists to this day. The pioneering work of William Froude on the flow resistance of ships, Ernst Mach on supersonic aerodynamics, Lord Rayleigh on hydrodynamic instability, Vincent Strouhal on excitation of oscillations by detached vortices, and Hermann von Helmholtz on vortex dynamics and scientific meteorology was followed by the groundbreaking research of Osborne Reynolds and Ludwig Prandtl, and later Andrei Nikolayevich Kolmogorov, which formed the basis of boundary layer and turbulence theory and in particular made an indispensable contribution to a deeper understanding of viscous flows in general, based on advanced mathematical methods. Considerable progress in solving the Navier–Stokes equation has been made since the middle of the 20th century thanks to the availability of computers and the development of efficient numerical methods. Thereafter, computational fluid dynamics (CFD) has emerged as an essential investigative tool in nearly every field of technology. Despite a well-developed mathematical theory and available commercial software codes, the computation of solutions of the governing equations of motion is still challenging, especially due to the nonlinearity involved, giving motivation for further research related to the mathematical and physical foundations. This book comprises seven peer-reviewed articles, four research articles, two reviews, and one technical report, covering a wide range of topics, methodical approaches, and their application to timely fluid flow problems. These include next to standard analytical and numerical methods also variational methods based on both deterministic and stochastic approaches. Next to incompressible flow problems like channel flow, vortex dynamics in turbulent flow, and flow through porous media, compressible flows are also addressed, including acoustic wave propagation in porous media. This volume will be of use as a reference to physicists, engineers, and mathematicians in both academia and industry.

Markus Scholle
Editor

Editorial

Physical and Mathematical Fluid Mechanics

Markus Scholle

ISAPS, Heilbronn University, D-74081 Heilbronn, Germany; markus.scholle@hs-heilbronn.de

Received: 31 July 2020; Accepted: 3 August 2020; Published: 5 August 2020

Abstract: Fluid mechanics has emerged as a basic concept for nearly every field of technology. Despite there being a well-developed mathematical theory and available commercial software codes, the computation of solutions of the governing equations of motion is still challenging, especially due to the nonlinearity involved, and there are still open questions regarding the underlying physics of fluid flow, especially with respect to the continuum hypothesis and thermodynamic local equilibrium. The aim of this Special Issue is to reference recent advances in the field of fluid mechanics both in terms of developing sophisticated mathematical methods for finding solutions of the equations of motion, on the one hand, and on novel approaches to the physical modelling beyond the continuum hypothesis and thermodynamic local equilibrium, on the other.

Keywords: analytical and numerical methods; variational calculus; deterministic and stochastic approaches; incompressible and compressible flow; continuum hypothesis; advanced mathematical methods

---

1. Introduction

Fluid Mechanics has a long history, going back at least to the era of ancient Greece, when Archimedes [1] investigated fluid statics and buoyancy and formulated his famous law, known now as Archimedes' principle, which was published in his work, "On Floating Bodies"—generally considered to be the first major work on fluid mechanics. Later, E. Torricelli and B. Pascal identified the pressure as a decisive physical quantity [2,3], while I. Newton [4] discovered the viscosity as another physical phenomenon of basic importance, which was later explored by J. L. M. Poiseuille and G. Hagen.

Mathematical fluid dynamics was first introduced by D. Bernoulli [5] and developed further by the mathematicians d'Alembert, Lagrange, Laplace, and Poisson, resulting in the well-known potential flow theory, being nowadays an essential topic in standard fluid dynamics text books [6–9]. Despite the obvious advantage of making various flow problems more tractable, the approach is restricted to inviscid and irrotational flows. A consistent mathematical treatment of viscosity by C.-L. Navier and G. G. Stokes led to the well-known Navier–Stokes equation, which, together with the continuity equation, continues to play the role of the essential field equation in fluid mechanics to this day. Initially, solutions of the Navier–Stokes equation could only be obtained for simple flow geometries until L. Prandtl discovered the mathematical singular boundary layer character of flows with high Reynolds numbers in the vicinity of rigid walls [10]. Prandtl's boundary layer theory and its advancement by T. von Kármán was a keystone both in a mathematical and a physical sense. Another branch of research is related to the formation of chaotic turbulent flow structures due to the nonlinearity of the Navier–Stokes equation, beginning with the early studies of O. Reynolds [11] and later advanced by G. I. Taylor [12] and A. Kolmogorov [13].

Considerable progress in solving the Navier–Stokes equation has been made since the middle of the 20th century, thanks to the availability of computers and the development of efficient numerical methods. Following this, computational fluid dynamics (CFD) has emerged as an essential investigative tool in nearly every field of technology. Despite there being a well-developed mathematical theory and available commercial software codes, the computation of solutions of the governing equations of motion is still challenging, especially due to the nonlinearity involved, giving motivation for further research related to the mathematical and physical foundations.

2. Overview of this Special Issue

Seven articles are published in the issue—four research articles, two reviews, and one technical report, covering a wide range of topics and methodical approaches.

In their research article [14], the formation of coherent vortex structures in a turbulent flow is analysed by direct numerical simulations, followed by image processing techniques and statistical analysis in order to identify and quantify streak characteristics of the flow. Motivated by the aim to complete our knowledge about and the understanding of vortices, the authors compare their findings to three standard vortex models, showing that they all give reasonably close results, and providing a deeper understanding of the interrelationships among different vortex models.

The basic mechanisms underpinning infiltration and drainage of water in soils and the role of viscosity is considered by Germann [15], introducing the basics of Newtonian shear flow in permeable media, presenting experimental applications and exploring the relationships of Newtonian shear flow with Darcy's law, Forchheimer's, and Richards' equations. An extension of the model to the transport of solutes and particles is finally presented.

Acoustic traveling waves in dual-phase media, such as a fluid in a porous solid, are investigated by Jordan [16], utilising the Rubin–Rosenau–Gottlieb theory of generalised continua. Exact and asymptotic expressions for linear and nonlinear poroacoustic waveforms are obtained. Numerical simulations are also presented, where von Neumann–Richtmyer "artificial" viscosity is used to derive an exact kink-type solution to the poroacoustic piston problem, and possible experimental tests of the findings are noted.

As a basic problem with respect to agricultural water resources, the turbulent flow in open channels is studied by [17], who derive a mathematical expression for the characteristic point location of depth average velocity in channels with flat or concave boundaries, particularly rectangular and semi-circular channels. For validation of the analytical model, experiments are carried out through comparison of velocity and discharge.

In their review article, [18] retrace alternative formulations of the Navier-Stokes equation based on potential fields, ranging from the classical potential theory to recent developments in this evergreen research field. The focus is centred on two major approaches which are diametrically opposed in their origin: (i) the Clebsch transformation originally applies to inviscid flow ($Re \to \infty$), while (ii) the classical complex variable method utilising Airy's stress function applies to Stokes' flow ($Re \to 0$). It is shown how both methods have been generalised by successive advancements and finally applied to the full Navier-Stokes equation, requiring the extension of the complex variable method to a tensor potential method. Basic questions relating to the existence and gauge freedoms of the potential fields and the satisfaction of the boundary conditions required for closure are addressed; with respect to (i), the properties of self-adjointness and Galilean invariance are of particular interest.

While most research in fluid mechanics is based on the continuum hypothesis, the stochastic variational description, based on the Lagrangian equations of motion in terms of material path lines instead of a field description, has proven to be a remarkable alternative to the classical theoretical, deterministic field approach. An obvious advantage of this approach is that it is very close to classical Newtonian mechanics, where the Lagrange formalism has been successfully established, allowing adoption of many of its features. It also closely refers to kinetic models in statistical physics. Cruzeiro [19] presents a selective review about this research field, regarding the velocity solving the deterministic Navier–Stokes equation as a mean time derivative taken over stochastic Lagrangian paths and obtaining the equations of motion as critical points of an associated stochastic action functional, involving the kinetic energy computed over random paths. Different related probabilistic methods are discussed.

Finally, the technical report by [20] analyses the damage characteristics and mechanisms of a disastrous groundwater inrush that occurred at the Luotuoshan coal mine on 1 March 2010, and gives a detailed overview about this incident in which 32 people lost their lives. The authors see a serious need for improvement in the timely detection of groundwater intrusion and its rapid rectification.

3. Conclusions

The seven publication contributions to the special edition cover a wide range of topics, provide valuable results, and point out open questions and possible future work. [14]'s analysis of the characteristic dimensions of streaky structures and vortices motivates the suggestion of straightforward hypotheses concerning the average width of streaks and the average distance between adjacent streaks, their development from the inner turbulent region to the outer region, the spanwise vortex density, and the coexistence of three different vortex structures as their contribution to improve the understanding of the mechanics of coherent structures in turbulent flows. [15] revealed novel aspects associated with Newton's infiltration that were not considerable in previous approaches to preferential flows, and state that the analytical expressions are amenable to mathematical procedures, such as kinematic wave theory, and their theoretical combinations may lead to new and solid hypotheses calling for experimental testing. Future work on the poroacoustic RRG theory is outlined by [16], who suggests the use of homogenisation methods in problems wherein the coefficients vary with position. Other possible extensions include the poroacoustic generalisation toward power-law fluids. Follow-on work might also include the study of poroacoustic signalling problems involving sinusoidal and/or shock input signals, as well as problems in which changes in entropy and temperature are taken into account. [17] consider an extension of future experimental studies of flow in open channels with regard to wall roughness to be very useful, especially with respect to the transition from smooth channels to vegetation-covered channels. Based on a detailed analysis and discourse, the two different potential approaches considered by [18] can be explained in the light of their different origins. Despite the very positive stage of development of both methods, some open questions remain, for instance, whether a general and all-encompassing potential approach exists, reducing to both the Clebsch and the tensor potential approach as special cases. The search for this "missing link" between two conceptually different approaches represents another future research topic of general interest. An extremely attractive further development of the tensor potential method would be the mapping of the entire problem to a matrix-algebra framework based on quaternions or Dirac matrices with the goal of developing highly efficient methods of solution. Having demonstrated the benefits of probabilistic methods for the study of the deterministic Navier–Stokes equation, Cruzeiro [19] envisages the development of novel numerical methods in the future. Finally, the tragic incident reported by Cui et al. [20] shows the need to detect and prevent such incidents in time with improved prediction models. Mathematical fluid mechanics can make a valuable contribution to this.

Funding: This research received no external funding.

Conflicts of Interest: The authors declare no conflict of interest.

References

1. Heath, T. *The Works of Archimedes*; Dover Books on Mathematics; Dover Publications: New York, NY, USA, 2002.
2. Driver, R.D. Torricelli's Law: An Ideal Example of an Elementary ODE. *Am. Math. Monthly* **1998**, *105*, 453–455.
3. Merriman, M. *Treatise on Hydraulics*; J. Wiley: Hoboken, NJ, USA , 1903.
4. Cohen, I. *Introduction to Newton's "Principia"*; History of Science; Harvard University Press: Cambridge, MA, USA, 1971.
5. Bernoulli, D. *Hydrodynamica, Sive, De viribus et Motibus Fluidorum Commentarii*; Dulsecker: London, UK, 1738.
6. Lamb, H. *Hydrodynamics*; Cambridge University Press: Cambridge, UK, 1974.
7. Panton, R.L. *Incompressible Flow*; John Wiley & Sons, Inc.: Hoboken, NJ, USA, 1996.

8. Batchelor, G.K. *An Introduction to Fluid Dynamics*; Cambridge Mathematical Library, Cambridge University Press: Cambridge, UK, 2000.
9. Spurk, J.H.; Aksel, N. *Fluid Mechanics*, 2nd ed.; Springer: Berlin, Germany, 2008.
10. Mayes, C.; Schlichting, H.; Krause, E.; Oertel, H.; Gersten, K. *Boundary-Layer Theory*; Physic and astronomy; Springer: Berlin, Germany, 2003.
11. Reynolds, O. III. An experimental investigation of the circumstances which determine whether the motion of water shall be direct or sinuous, and of the law of resistance in parallel channels. *Proc. R. Soc. Lon.* **1883**, *35*, 84–99, doi:10.1098/rspl.1883.0018.
12. Batchelor, G.K. Geoffrey Ingram Taylor, 7 March 1886–27 June 1975. *Biograph. Mem. Fell. R. Soc.* **1976**, *22*, 565–633. [CrossRef]
13. Kolmogorov, A.N. The local structure of turbulence in incompressible viscous fluid for very large Reynolds numbers. *Proc. R. Soc. Lond. Math. Phys. Eng. Sci.* **1991**, *434*, 9–13. [CrossRef]
14. Wang, H.; Peng, G.; Chen, M.; Fan, J. Analysis of the Interconnections between Classic Vortex Models of Coherent Structures Based on DNS Data. *Water* **2019**, *11*, 2005. [CrossRef]
15. Germann, P. Viscosity Controls Rapid Infiltration and Drainage, Not the Macropores. *Water* **2020**, *12*, 337. [CrossRef]
16. Jordan, P.M. Poroacoustic Traveling Waves under the Rubin–Rosenau–Gottlieb Theory of Generalized Continua. *Water* **2020**, *12*, 807. [CrossRef]
17. Li, T.; Chen, J.; Han, Y.; Ma, Z.; Wu, J. Study on the Characteristic Point Location of Depth Average Velocity in Smooth Open Channels: Applied to Channels with Flat or Concave Boundaries. *Water* **2020**, *12*, 430. [CrossRef]
18. Scholle, M.; Marner, F.; Gaskell, P.H. Potential Fields in Fluid Mechanics: A Review of Two Classical Approaches and Related Recent Advances. *Water* **2020**, *12*, 1241. [CrossRef]
19. Cruzeiro, A.B. Stochastic Approaches to Deterministic Fluid Dynamics: A Selective Review. *Water* **2020**, *12*, 864. [CrossRef]
20. Cui, F.; Wu, Q.; Xiong, C.; Chen, X.; Meng, F.; Peng, J. Damage Characteristics and Mechanism of a 2010 Disastrous Groundwater Inrush Occurred at the Luotuoshan Coalmine in Wuhai, Inner Mongolia, China. *Water* **2020**, *12*, 655. [CrossRef]

Article

Analysis of the Interconnections between Classic Vortex Models of Coherent Structures Based on DNS Data

Hao Wang *, Guoping Peng, Ming Chen and Jieling Fan

College of Civil Engineering, Fuzhou University, Fuzhou 350116, China; pengguoping_1314@163.com (G.P.); korry_chen@163.com (M.C.); fanjieling318@163.com (J.F.)

* Correspondence: wanghao_0512@163.com

Received: 7 August 2019; Accepted: 23 September 2019; Published: 26 September 2019

Abstract: Low- and high-speed streaks (ejection, Q2, and sweep, Q4, events in quadrant analysis) are significant features of coherent structures in turbulent flow. Streak formation is closely related to turbulent structures in several vortex models, such as attached eddy models, streamwise vortex analysis models, and hairpin vortex models, which are all standard models. Vortex models are complex, whereby the relationships among the different vortex models are unclear; thus, further studies are still needed to complete our understanding of vortices. In this study, 30 sets of direct numerical simulation (DNS) data were obtained to analyze the mechanics of the formation of coherent structures. Image processing techniques and statistical analysis were used to identify and quantify streak characteristics. We used a method of vortex recognition to extract spanwise vortices in the x–z plane. Analysis of the interactions among coherent structures showed that the three standard vortex models all gave reasonably close results. The attached eddy vortex model provides a good explanation of the linear dimensions of streaky structures with respect to the water depth and Q2 and Q4 events, whereby it can be augmented to form the quasi-streamwise vortex model. The legs of a hairpin vortex envelop low-speed streaky structures and so move in the streamwise direction; lower-velocity vortex legs also gradually accumulate into a streamwise vortex. Statistical analysis allowed us to combine our present results with some previous research results to propose a mechanism for the formation of streaky structures. This study provides a deeper understanding of the interrelationships among different vortex models.

Keywords: image processing; streaky structures; hairpin vortex; attached-eddy vortex; streamwise vortex

---

1. Introduction

Turbulence is generally not altogether chaotic, whereas there are many regular coherent structures in a turbulent flow. The coherent structures include streaky structures formed by low- and high-speed streaks, the bursting phenomenon that includes ejection and sweep events (in quadrants Q2 and Q4), vortex structure models (streamwise vortex model, attached eddy vortex model, hairpin vortex model and hairpin vortex groups), as well as superscale structures.

Low- and high-speed streaks are important in turbulence dynamics because of their large scale [1]. Experimental research into low- and high-speed streaks using hydrogen bubbles was first conducted by Kline et al. [2]. The characteristic scales of streaky structures were also identified by many researchers as the average nondimensional width W = 20–40y_* and spanwise distance D = 100y_* in the boundary layer region [3–5]. Note that $y_* = v/u_*$ defines the inner scale, where v is kinematic viscosity and u^* is friction velocity, which represents the shear stress velocity. Lin et al. [6] used particle image velocimetry

(PIV) to capture the flow fields. Their results show that the spatial distribution of high-speed streaks is similar to that of low-speed streaks.

Zhong et al. [1] identified elongated streamwise low- and high-speed streaks near the free surface in open-channel flows by spanwise correlation analysis. The presence of large-scale streaks across the whole flow depth has been confirmed by many researchers. Previous evidence indicates that the distance between neighboring low-speed streaks is the water depth scale (H–$2H$) [7–10]. Sukhodolov et al. [11] found that streamwise streak length could exceed $3H$ while Zhong et al. [1] found the length to be greater than $10H$. The existence of streaky structures throughout the whole turbulent layer is now commonly accepted [7,12–14].

Various hypotheses and models of vortices have been created to explain the formation of low- and high-speed streaky structures. Many studies proposed super-streamwise vortex models of Q2/Q4 events, which included alternating low- and high-speed streaks in the spanwise direction [9,15,16]. The attached eddy hypothesis developed by Townsend [17] explained Q2/Q4 events and the development of streaky structures, which scale linearly with their water depth from the inner region to the outer region. Adrian and Marusic [18] advocated a model using hairpin vortices and packets: hairpins and packets cause the ejection of low-speed streaks between the two legs of the hairpin vortex when the super-streamwise vortices feed themselves by sweeping low-momentum hairpins and packets into the low-speed regions. Secondary flow cells have also been modelled as vortices which originate in the vicinity of the side walls [19,20].

The existing research indicates that vortex models have limited use. Researchers accept the existence of super-streamwise vortices theoretically, but the literature reviewed above shows that there is no consensus among researchers concerning the formation of vortices. For example, the streamwise vortex model cannot explain how streak length varies linearly from inner region to outer region. The attached eddy vortex is a single structure, which does not explain the distribution and organization of the many funnel vortices in turbulent flow. Hairpin vortex models are usually developed for a single flow field and vortex structure in the x–y or x–z plane. However, current understanding of the characteristics of hairpin vortices is insufficient to generate a robust interpretational theory. There are relatively few studies of vortex models, and thus there is a lack of systematic quantitative vortex model analysis; vortex models can still be improved.

We used models to investigate vortices as coherent structures in turbulent flow, using direct numerical simulation (DNS) data. We identified the positions of low- and high-speed streaks using image processing and calculated the characteristic dimensions of streaky structures in both the inner and outer layers using a statistical method. We identified streamwise vortices, attached eddy vortices, and hairpin vortices by analyzing the variation in streak dimensions with respect to water depth and analyzed the spatial relationships between streaky structures and spanwise vortex position to explain the relationship between the three vortex models. Finally, we propose a new hypothesis.

The remainder of this paper is organized as follows. Section 2 describes the methods used to analyze the DNS data and to identify and calculate the characteristic dimensions of streaky structures. Section 3 presents an analysis of the regular variation in streaky structures and the mechanics of the three vortex models. Section 4 offers a summary and a brief discussion of our major findings and the conclusions we draw from them.

2. Materials and Methods

2.1. Closed Channel Flow: DNS

Particle image velocimetry (PIV) is the principal experimental method of measuring the flow field. The area captured by the camera is relatively small, due to the limited intensity of laser light, as the physical width (z direction) of the image. Thus the number of low- and high-speed streaks sampled is relatively small, whereby the characteristic scale of streaks is not particularly accurate. Therefore, we used the numerical data of Del Alamo et al. [21] to investigate coherent structures in turbulent flow, and thereby obtained complete flow field information for closed channel flow. Data series L950, which we used extensively, contains data for almost all recognized large-scale coherent structures scaled by water depth [5,21,22]. Figure 1 shows that the dimensionless length and width of the DNS flow field are both large, whereby the characteristic scales of the streaks are relatively accurate. We give a brief introductory summary here. Detailed information can be found in Del Alamo et al. [21].

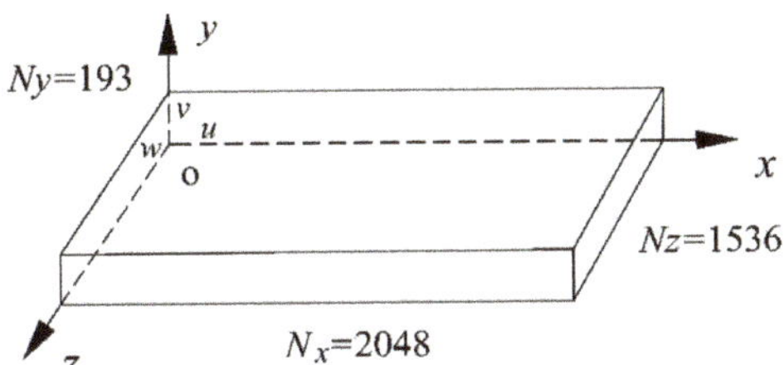

Figure 1. The computation region of direct numerical simulation (DNS) in closed channel flow.

The friction Reynolds number for the flow was 934, which indicates that the range of temporal and spatial fluid scales involved in turbulence was considered to be relatively large. The simulation covers a spatial domain (x, y, z) of $16\pi h/3 \times 1h \times 2\pi h$, where h is the half-channel height and the domain is discretized into an array $(x \times y \times z)$ of $2048 \times 193 \times 1536$ points. Each grid point contains three velocity components corresponding to nine velocity gradient data points. The streamwise (x), vertical (y), and spanwise (z) dimensions are shown in Figure 1, which summarizes of the DNS data; u, v, and w represent the instantaneous velocities in the x, y, and z directions, respectively. Major parameters of the DNS data are summarized in Table 1.

Table 1. Parameters of the DNS (data from Del Alamo et al. [21]).

Parameter	L_x/H	L_z/H	L_y/H	Δx^+	Δz^+	Δyc^+	N_x	N_z	N_y
Original	8π	3π	2	7.6	3.8	7.6	3072	2304	385
Present study	$16\pi/3$	2π	1	7.6	3.8	7.6	2048	1536	193

In Table 1, H is the water depth; L_x, L_y, and L_z are the spatial domains along the x, y and z directions, respectively; Δx and Δz are the grid resolutions in the x and z directions, respectively; N_x and N_z correspond to the grid numbers; Δy_c is wall-normal grid spacing at the channel center; N_y represents the grid numbers along the y direction; the superscript $^+$ denotes normalization by the inner scale (u^* and v); and u^* is the friction velocity and represents the shear stress velocity, for example, $\Delta x^+ = \Delta x u^*/v$.

For each of the three-dimensional instantaneous velocity fields, totals of 153×30 x–y planes, 204×30 y–z planes, and 193×30 x–z planes were extracted for analysis. There were 2048×124 (x–y plane), 245×1536 (y–z plane), and 2048×1536 (x–z plane) grid points. We extracted 193×30 x–z planes for analysis, and there are 2048×1536 grid points in each x–z plane.

2.2. Detection of Streaky Structures

The formation of low- and high-speed streaks are related to instantaneous turbulence fluctuations. Three steps were followed to study the characteristics scale of streaky structures: (1) the detection function was used to identify the high- and low-speed streaks; (2) image processing, including binarization and morphological operations, was used to extract the image structure of both low- and high-speed streaks [6,10]; and (3) statistical analysis was used to calculate the characteristic scales of streaks.

2.2.1. Detection Function

The method, after modification, used the following two functions,

$$F_d(m,n,y^+,t) = \frac{u'(m,n,y^+,t)}{u_{std}(y^+)} \quad (1)$$

$$Ct(y^+) = C \times \max[u_{std}]/u_{std}(y^+) \quad (2)$$

where (m, n) is the grid position in the x–z plane; u' is the streamwise velocity fluctuation; $u_{std}(y^+)$ is the standard deviation of the streamwise velocity at y^+; $C_t(y^+)$ is the water depth threshold at y^+; F_d is the dimensionless value of detection function; C is a constant, equal to 0.6, as recommended by Lin et al. [6]; and $\max[u_{std}]$ is the maximum value of u_{std} in the flow domain. $F_d > Ct$ (high-speed) and $F_d < -Ct$ (low-speed) identify the streaks. Justification for the two equations and specific details are provided in Wang et al. [10].

Figure 2a shows the contours of F_d for low-speed streaks at $y^+ = 21.05$. The positive and negative values of F_d indicate the existence of instantaneous streamline fluctuations, forming the low- and high-speed streaks. Low-speed regions (brown), high-speed regions (blue), and other flow regions (green) can be recognized by applying a threshold value of $C_t(y^+)$ to the contour map, as shown in Figure 2b.

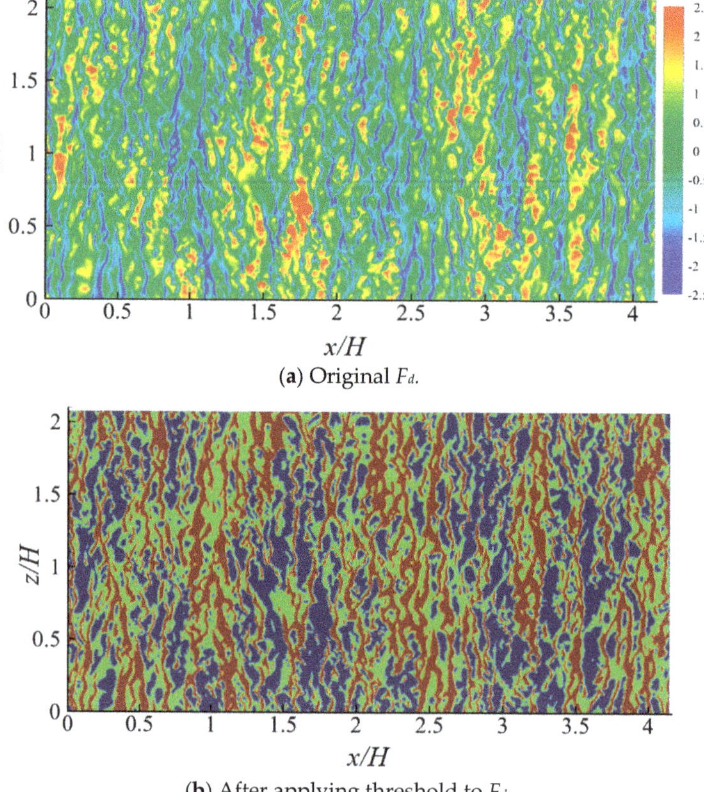

(**a**) Original F_d.

(**b**) After applying threshold to F_d.

Figure 2. Visualization of streaks represented by the dimensionless value of detection function F_d: (**a**) original F_d with the range of the color bar set from −2.5 to 2.5; (**b**) after applying the threshold value to F_d.

2.2.2. Image Processing

To better quantitatively analyze the low- and high-speed streaks, a binary procedure was used to extract the streaks: values less than $-C_t$ were assigned the value 1, whereas values greater than $-C_t$ were assigned a value 0. Figure 3 shows the image processing procedure for extracting low-speed streaks. The procedure for extracting high-speed streaks is similar, but uses a different C_t threshold value.

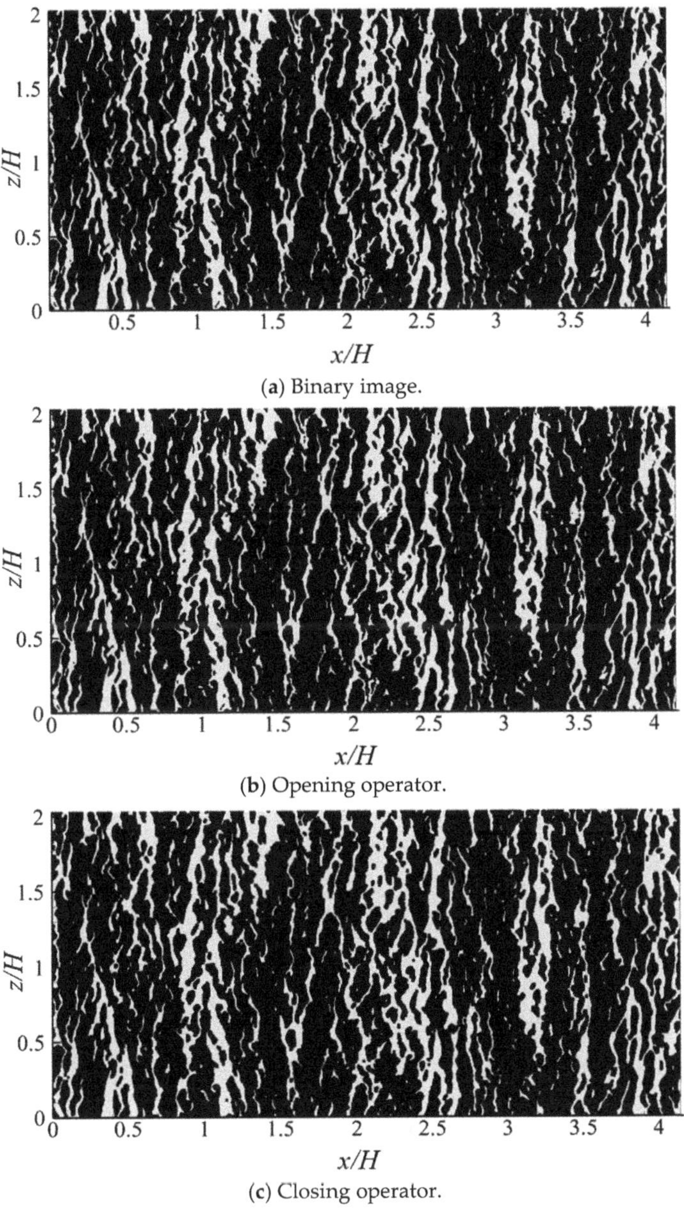

(**a**) Binary image.

(**b**) Opening operator.

(**c**) Closing operator.

Figure 3. *Cont.*

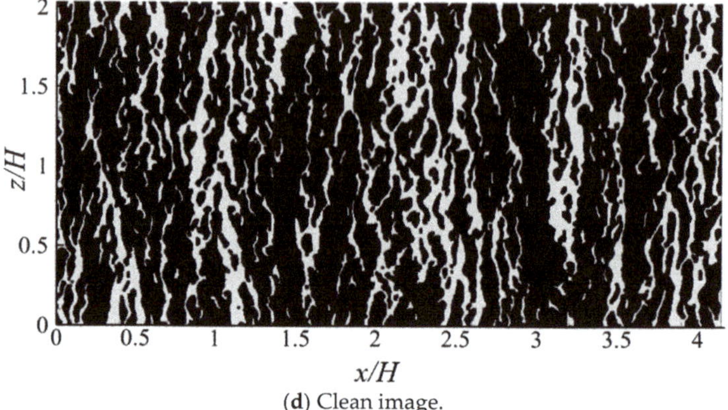

(**d**) Clean image.

Figure 3. Image processing: (**a**) binary image, (**b**) opening operator, (**c**) closing operator, and (**d**) clean image.

The original F_d image was binarized (Figure 3a), and a basic morphological transformation was used to filter out noise in the binary images. This transformation was done in two steps [23]. First, the opening operator (Figure 3b) and closing operator (Figure 3c) were used to delete some isolated regions and fill some holes. The opening operator is derived from the fundamental morphological operations of dilation as well as erosion and was used to break the adhesion between objects and remove small particle noise; the closing operator combines the operations of erosion and dilation and can be used to connect neighboring regions and fill in small holes. The area of the streak graph does not change significantly during calculation when using the opening and closing operators.

Second, some isolated objects were deleted from the binary image with the *bwareaopen* function. After these two steps were performed, the streaky structures were clearly visible (Figure 3d). The selection of specific parameters and values is described by Lin et al. and Wang et al. [6,10].

2.2.3. Model of Streaky Structures

We used streak width (w) and the distance between adjacent streaks (d) to scale and characterize streaky structures. As the structures vary spatiotemporally, the image was parsed line-by-line to quantify both w and d. We assumed that the number of streaky structures in a line of the image was ns. The streak widths are denoted by $w_1, \ldots, w_{i-1}, w_i, w_{i+1}, \ldots, w_{ns}$ when the streak distances are denoted by $d_1, \ldots, d_{i-1}, d_i, d_{i+1}, \ldots, d_{ns-1}$.

For the random row in the image (rth line), the mean streak width of the rth line, $w_{(r)}$, and the mean streak width of the whole velocity field at each y_+, $W_{(y+)}$, were obtained by Equations (3) and (4):

$$w_{(r)} = \frac{\sum_{i=1}^{ns} w_i}{ns}, (r = 1, 2, \ldots, m) \qquad W_{y^+} = \frac{\sum_{r=1}^{m} w_{(r)}}{m} \tag{3}$$

where m is the total number of rows of the flow field. Similarly, the mean spanwise distance of each row, $d_{(r)}$, and the mean spanwise distance of the whole velocity field at each y^+, $D_{(y^+)}$, were calculated by

$$d_{(r)} = \frac{\sum_{i=1}^{ns-1} (d_{i+1} - d_i)}{ns - 1}, (r = 1, 2, \ldots, m) \qquad D_{(y^+)} = \frac{\sum_{r=1}^{m} d_{(r)}}{m} \tag{4}$$

The dimensionless characteristic scales of the streaks can be obtained by $D^+ = Du^*/v$ and $W^+ = Wu^*/v$. The 30 instantaneous x–z velocity fields (DNS data) were captured at each y position in all cases.

Comparisons between characteristic scales and previous data are given in Figure 8 of Wang et al. [10]. The variations in the mean spanwise distance relative to calculated wall distance are completely feasible, and the above method can be used to analyze the characteristic scales of streaky structures.

3. Results

The relationship between the spanwise distance between streaks and the vortex model used is significant for analysis of the entire phenomenon. The spanwise interstreak distances for both low- and high-speed streaks *D/H* and water depth *y/H* were plotted. Figure 4 shows that the trend of high-speed streaks is similar to that of low-speed streaks. As water depth *y/H* increases, *D/H* reaches a turning point close to half the water depth of the closed-channel flow. The increased amplitude decreases significantly near the half water depth due to a weak boundary layer.

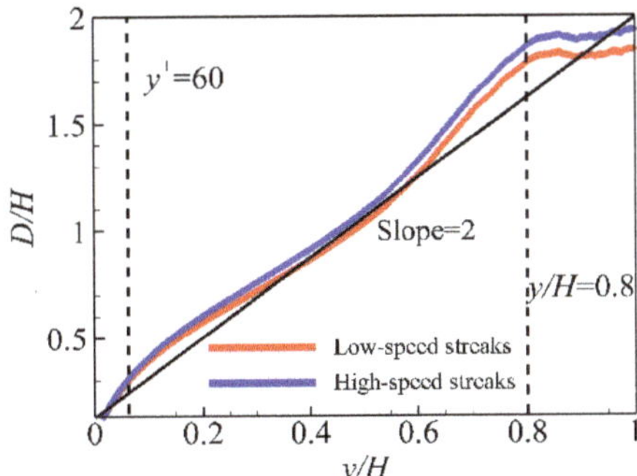

Figure 4. Spanwise distances of streaks; *D/H* varies with water depth *y/H*.

Most research on streaky structures has been concerned with the inner region (y^+) and outer region (*H*) scales. The relationship between two streaky structures of different scales is unclear. We calculated the spanwise distances over the entire water depth continuously for both the inner and outer regions. The results show that the development of streaky structures along the entire flow depth is a continuous process.

D/H increases linearly with *y/H* in the outer layer (i.e., when $0.1 < y/H < 0.8$), and the slope is approximately 2. Streaky structures are closely linked to the vortex model used. Our results are interpreted in the context of the streamwise and attached eddy vortex models as follows.

3.1. Streamwise Vortex Model

The spanwise distances are approximately twice the water depth, and the formation of streaky structures in the outer layer is related only to water depth. This is consistent with streamwise vortex structures being generated automatically from the self-organization of wall-bound turbulence [7]. In this case, the streamwise vortex also shows that the strong pumping action of low-speed fluid creates an ejection event in the associated second quadrant, Q2 ($u < 0, v > 0$). Fluid moving at high speed from the water surface toward the bed creates a sweep event in Q4 ($v < 0, u > 0$). The low- and high-speed streaks are located at the downwelling and upwelling sides of the streamwise vortices respectively.

The conceptual streamwise model in Figure 5 was built to explain the formation of streaky structures and Q2/Q4 events; the distance between adjacent streaks near the water surface is twice the water depth [1,14,18].

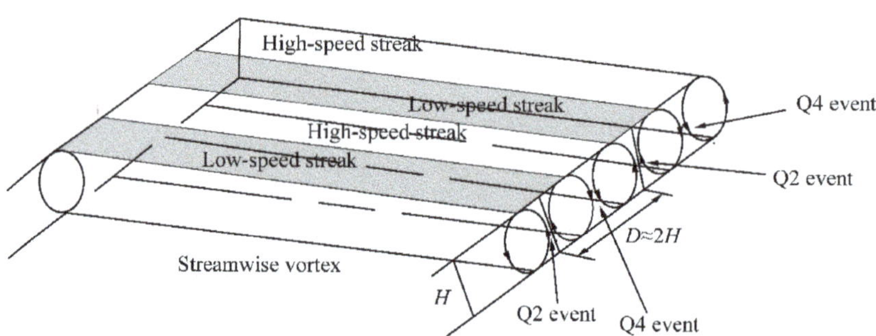

Figure 5. Representation of the streamwise vortex model.

3.2. *Attached Eddy Vortex Model*

Figure 6a shows that the increases in the characteristic scales of the streaks are linear, which is a core assumption of Townsend's attached eddy vortex hypothesis [17,23]. Our results indicate the suitability of the attached eddy vortex model.

Figure 6a shows that, according to the model, the attached eddy vortex develops from the near-wall region into a conical vortex in the streamflow direction. The formation mechanics of low- and high-speed streaks and the Q2/Q4 events are similar to those of the streamwise vortex. In particular, the spanwise distances between adjacent low-speed (or high-speed) streaks are closely related to the size of the streamwise vortices and thus linearly proportional to y. Three cross-stream plane sections (*slices*) of the attached eddies at different water depths y are shown in Figure 6; the blue and red backgrounds indicate the low-speed and high-speed streaks, respectively. Figure 6a shows the Q2\Q4 events when Figure 6b shows that the dimensions of the vortex increase linearly as water depth increases and the relationship between D and h remains basically constant with $D \approx 2h$. Overall, the attached eddy vortex model explains the distance between adjacent streaks from the inner region to the outer region.

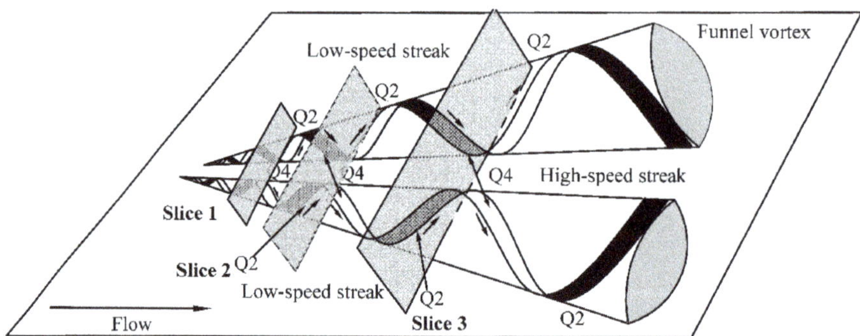

(**a**) Diagram of the streaky structures and a pair of attached-eddy hypothesis.

Figure 6. *Cont.*

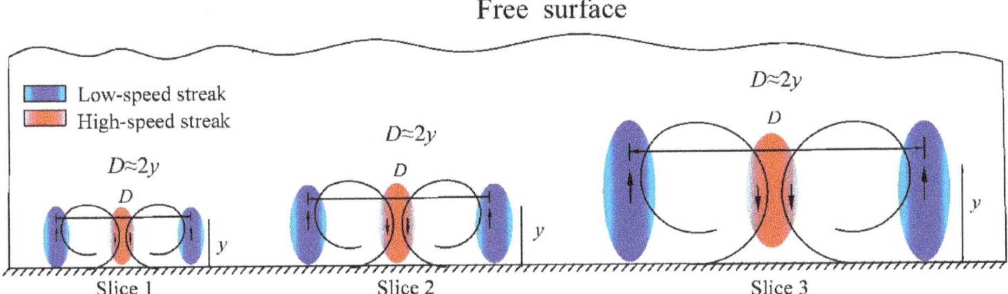

(**b**) High- and low-speed streaky structures for each slice.

Figure 6. Streaky structures described in terms of the attached eddy hypothesis: (**a**) A pair of vortices and the (**b**) corresponding high- and low-speed streaky structures for each slice.

3.3. Hairpin Vortex Model

The hairpin vortex model developed by Adrian is a relatively new vortex model [24]. We investigated the positional relationships between spanwise vortices and streaky structures in the x–z plane to determine the suitability of the hairpin vortex model.

3.3.1. Vortex Extraction in the X–Z Plane

We used the two-dimensional swirling-strength λ_{ci}-criterion [25]. Streaky structures form in the in x–z plane, so a brief introduction to extracting a vortex in the x–z plane is given.

The swirling strength λ_{ci} is given by

$$\lambda_{ci} = \begin{cases} \sqrt{R - P^2/4}, & 4R - P^2 > 0\,; \\ 0 & 4R - P^2 \le 0\,; \end{cases} \tag{5}$$

where

$$P = -\frac{\partial u}{\partial x} - \frac{\partial w}{\partial z}, R = \frac{\partial u}{\partial x}\frac{\partial w}{\partial z} - \frac{\partial u}{\partial y}\frac{\partial w}{\partial z} \tag{6}$$

Following Wu and Christensen [26], we defined the normalized swirling strength Λ_{ci} as $\Lambda_{ci} = \lambda_{ci}\omega_z/|\omega_z|$, where ω_z is the fluctuating spanwise vorticity and λ_{ci} and Λ_{ci} are swirling strength discriminators. $\Lambda_{ci}^{rms}(y)$ is the local root mean square of Λ_{ci} at the wall-normal position y, and we defined the normalized swirling strength Ω_{ci} by

$$\Omega_{ci}(x, y) = \frac{\Lambda_{ci}(x, y)}{\Lambda_{ci}^{rms}(y)} \tag{7}$$

In an ideal fluid, there is a clear boundary between rotating and irrotational fluid. Zero (0) can be used as a threshold to easily extract the vortex. However, in a nonideal (actual) fluid, viscosity causes dissipation of the vortex, which greatly complicates vortex identification. We used a non-zero threshold of 1.5 to identify a vortex, following the recommendation of Wu and Christensen [26], so that

$$|\Omega_{ci}| \ge 1.5 \tag{8}$$

Negative or positive values of Ω_{ci} in Equation (8) represent a clockwise or counterclockwise vortex, respectively.

3.3.2. Vortex Density

When the vortex structure has been determined by the preceding methods, the vortex population density Π^+ is calculated by

$$\Pi^+ = \frac{N_{vortex}(y^+)}{Nx(y^+) \cdot Nz(y^+) \cdot \Delta x^+ \cdot \Delta z^+} \quad (9)$$

where N_{vortex} is the spanwise number of vortices at position y^+.

The prograde and retrograde vortices in the x–z plane were separated, and the population densities of the vortices were computed for each position of y^+. Figure 7 shows that the population density of 2D vortices varies with water depth, reaching a maximum in the near-wall region at $y^+ = 40.22$. This result may be partly due to the number of streamwise vortices in the deeper water. In the outer region, vortex density decreases gradually as y^+ increases. As the shear stress in the x–z plane is approximately zero, the population density of prograde vortices is equal to that of retrograde vortices at each y^+ position. These results agree with those obtained by Chen et al. [27], which confirms the logic of our vortex extraction method.

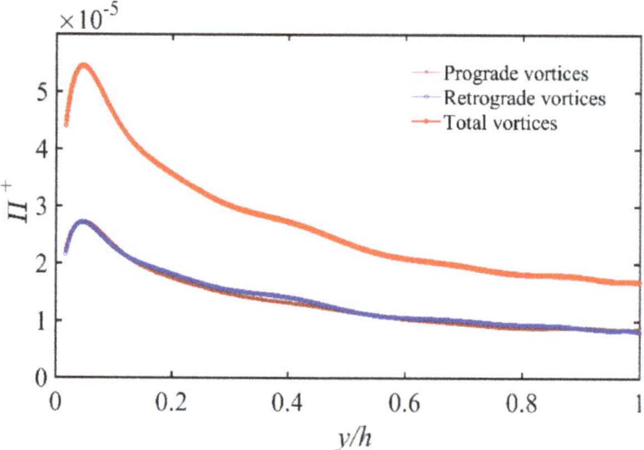

Figure 7. Vortex population density in the x–z plane.

3.3.3. Location of Vortices and Streaks

Figure 8 shows the cores of spanwise vortices surrounded by nine velocity vectors (red), high-speed streaks (yellow), low-speed streaks (blue), and the in-between region (green) that were obtained by the preceding methods. We extracted and analyzed 400 × 800 grid points in the x–z plane. The dimensionless area is 3040 × 3040, and x^+ and z^+ represent the dimensionless length along the streamwise and spanwise directions, respectively.

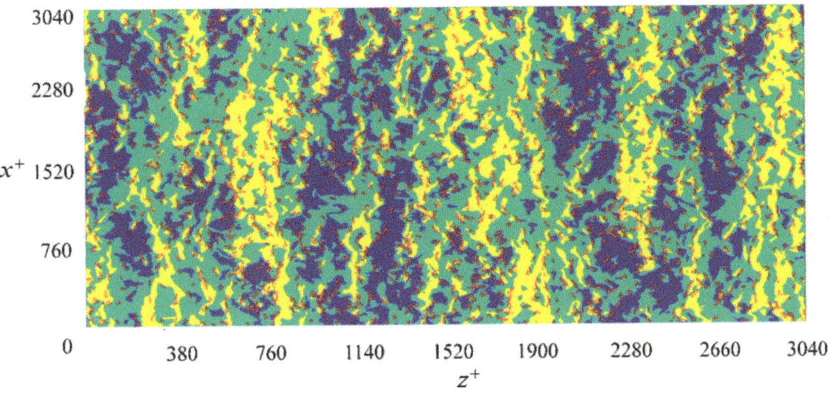

Figure 8. Positional distributions of streaks and spanwise vortices.

To further investigate the relationship between streaky structures and spanwise vortices, the vortex core ($u_{i,j}$) velocity and the eight surrounding streamwise velocities (Figure 9) were averaged using Equation (10). Equation (11) was then used to calculate F_d at the core of each vortex. We use $C_t(y)$ to identify the region of the distribution of vortex cores.

$$\overline{u} = \frac{1}{9}\sum_{i=1}^{9} u_i \tag{10}$$

$$\overline{F_d} = \frac{\overline{u}'}{u_{std}(y^+)} \tag{11}$$

where $\bar{u}$ is the average velocity of spanwise vortices in the x–z plane, $\bar{u}'$ is the fluctuating velocity, and $\overline{F_d}$ is the detection function of average vortex velocity. We can now calculate the statistical measures of the vortices in different streaky structures.

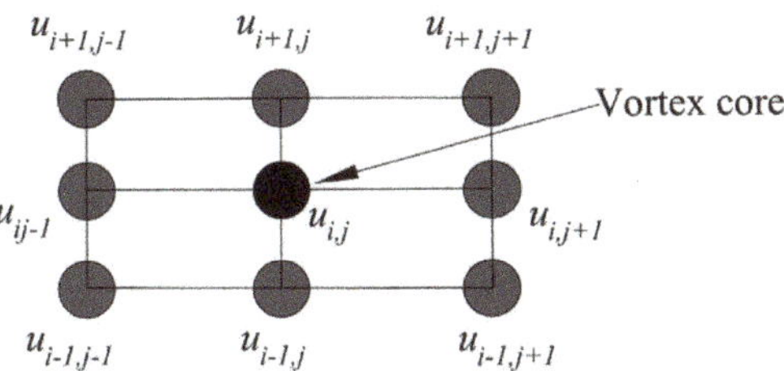

Figure 9. Grid for streamwise velocity u showing the core of the vortex.

In previous research, streaks have generally been divided into low- and high-speed. However, to obtain a more detailed analysis of the relationships between spanwise vortices and streaky structures, we divided the x–z plane into three types using the threshold $C_t(y^+)$: low-speed streaks, high-speed streaks, and in-between regions.

Figure 10 shows that the numbers of vortices differ greatly within the different streaks. The spanwise vortices in the in-between region occur in the greatest numbers, followed by low-speed streaks, with the least numbers in the high-speed streaks. When $y/H > 0.1$ the number of vortices in high-speed streaks is approximately equal to 0.

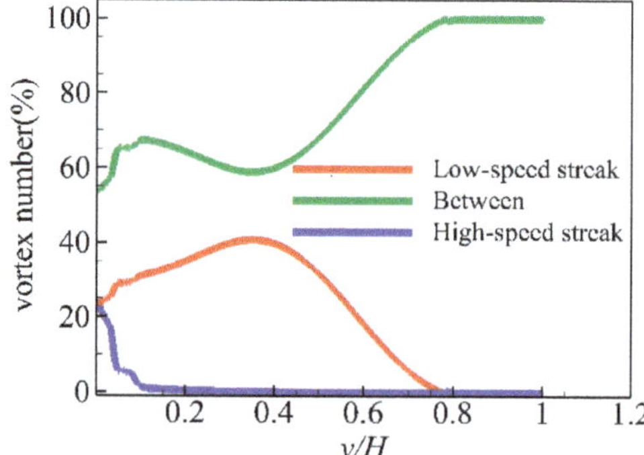

Figure 10. Vortex numbers in low- and high-speed streaks and in-between regions.

It should be noted that the three areas occupied by the three types are also different. By dividing the number of vortices by the area containing them, we eliminated the influence of area from our analysis to better understand the distributions of spanwise vortices located in streaky structures.

3.3.4. Vortex Density in Different Streaks

We first calculated the areas of the low- and high-speed streaks. The values of both low- and high-speed streak widths are influenced by the threshold value (C_t) and are difficult to recognize relative to the spanwise distance. If C_t is too large, the streak width will not include the whole with of the streak; nevertheless, if C_t is too small, the streak width will contain parts of the in-between region. However, even with different threshold values, the change tendencies are basically consistent. Here, we used the threshold value suggested by Lin et al. [6] to identify the width of low- and high-speed streaks. Figure 11 shows that as y^+ increases, streak width first increases and then decreases. As the water depth increases from inner region to outer region, the streak scale also increases. When water depth is close to the surface (about 0.7H), the weak boundary layer restrains the streak scale, and the streak width will decrease. This trend is stable and clearly demonstrated.

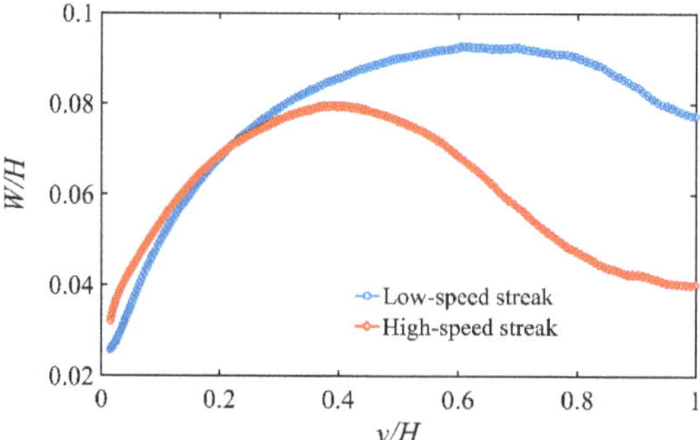

Figure 11. Characteristic dimensions (width) of low- and high-speed streaks varying with y^+.

The area percentages of low- and high-speed streaks in the x–z plane at each position y+ are also important characteristic dimensions and can be regarded as the normalized areas of streaks. The area percentage, defined as P_s, can be obtained by

$$A_s = \sum_{i=1}^{m} ns_i \cdot w_i \tag{12}$$

$$P_s = \frac{A_s}{A_t} \times 100\% \tag{13}$$

where A_s is the total area of low- or high-speed streaks and A_t is the total area of the flow field (2048 × 1536). Figure 12 shows that the percentage area of both low- and high-speed streaks decreases as y^+ increases. In the near-wall region ($y/h < 0.1$), the gradients are steep; in the outer region ($y/h > 0.3$), the decreasing trends become less steep. This result shows that the streaks occur mainly in the near-wall region where there is high shear stress. Mean shear stress decreases as y/h increases, and so does its effect on the streaks. Streaks in the outer region decrease in number and so the percentages of both low- and high-speed streaks in the x–z plane also decrease.

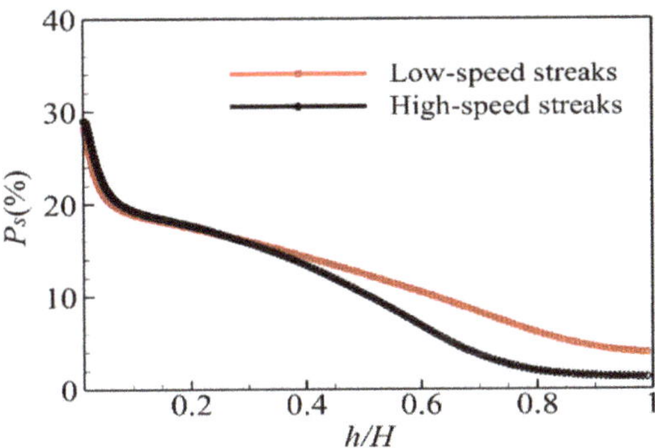

Figure 12. Percentage area of low- and high-speed streaks along the x–z plane.

Figures 11 and 12 both show that the formation of streaks between the near-wall region and the outer layer is a continuous process, as also shown in Figure 4.

3.3.5. Calculation of Vortex Density

Equation (10) was used to calculate the density of spanwise vortices in different streaky structures, as shown in Figure 13. The density of spanwise vortices is highest in low-speed streaks, intermediate in the in-between region and least in the high-speed streaks. Thus, there are big differences between the number of vortices and vortex density located in different streaky structures.

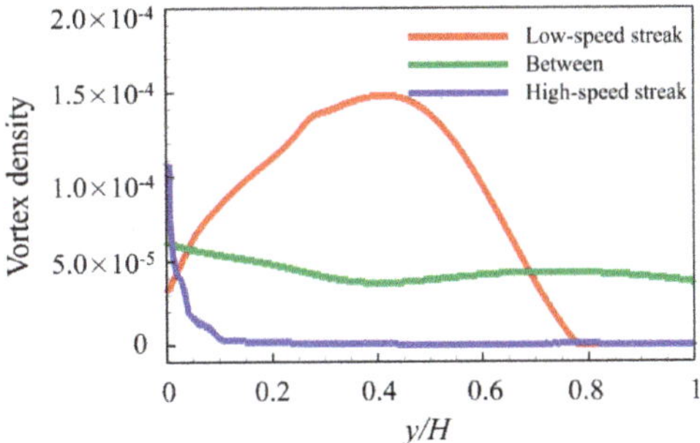

Figure 13. Vortex densities located in different streaky areas.

The hairpin vortex model developed by Adrian [24] has gained widespread acceptance, due to experimental visualization using particle image velocimetry and direct numerical simulation. Figure 14 shows the standard coherent structure model of a hairpin vortex developed by Adrian [24]. The alignment of coherent vortices induces a low-speed fluid region inside the hairpin packets.

Due to the closed-loop feedback cycle between hairpin vortex cells and streamwise vortices [1,28], the streamwise vortices are stable.

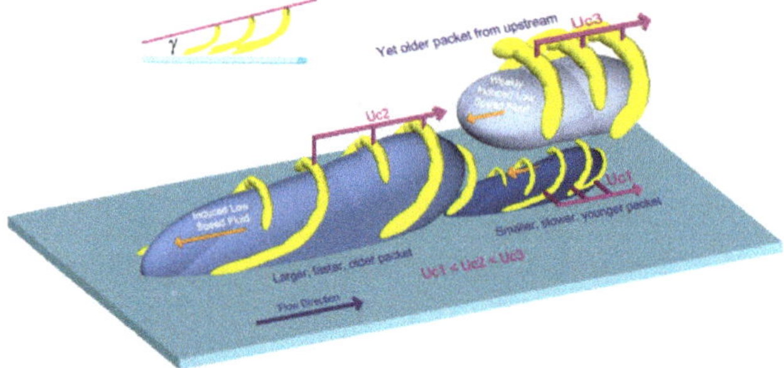

Figure 14. Representation of hairpins and hairpin packets by Adrian [24].

Research into hairpin vortex behavior has become an important direction of research. However, the sample sizes used in the research are fairly small, so the regularity of the relationship between streaky structures and spanwise vortices in the x–z plane must be further researched by analyzing large samples.

We obtained statistics from large DNS data samples, and we found that spanwise vortex density in low-speed streaks is greater than in high-speed streaks. This result indicates that hairpin vortex legs are closer to the low-speed streaks and further from the high-speed streaks. Thus, the results we obtained exhibit an important feature of hairpin vortex legs when they envelop low-speed streaks to move along the quasi-streamwise direction, as shown in Figure 14. The legs of the hairpin vortex are spanwise vortices in the x–z plane, as shown in Figure 15. Spanwise vortices are mainly distributed in the region of low-speed streaks, consistent with the structure of hairpin vortices. Our results support the logic of the hairpin vortex model and reveal mechanisms of hairpin vortex behavior more explicitly.

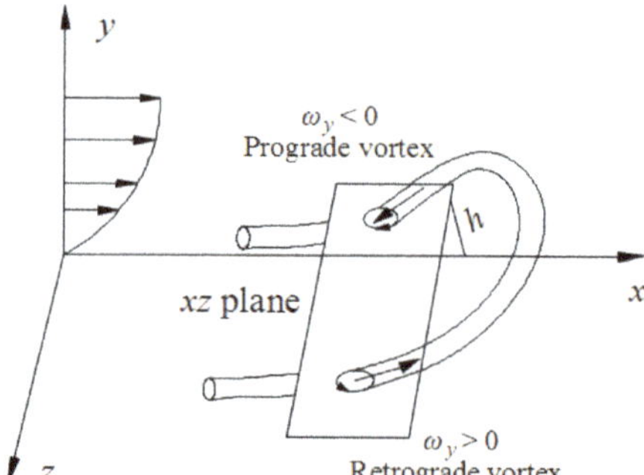

Figure 15. Representation of hairpin vortex and spanwise vortex.

4. Discussion and Conclusions

4.1. Discussion

A large-scale vortex is a conceptual model, or representation, of a natural phenomenon intended to be used in the provision of logical explanations of all kinds of coherent structures in turbulent fluids. Classical and prevailing views of vortices have led to many vortex models being developed. We identified both low- and high-speed streaks from the wall to the surface using image processing technology; the meandering large scale motions are impossible to ignore. The low- and high-speed streaks are formed by an ejection event (Q2, $u < 0$, $v > 0$) and a sweep event (Q4, $v < 0$, $u > 0$) [14,18].

We used the super-streamwise vortex (Figure 5) as the interpretative model to explain the preceding results and the spanwise distance of nearby streaks (2*H*). We found that the scale of the streaks increased in proportion to their distance from the wall. The result is consistent with the classical model, which combines length growth with growth in eddies, developed by Townsend [17]. Our results also explain the logarithmic growth in open channel flow. The distributions of spanwise vortex density in low- and high-speed streaky structures suggest further research into hairpin vortices. We statistically sampled large datasets to compare and analyze three vortex models. Our analysis of the results shows the benefit of explaining coherent structures from the three different model perspectives.

The literature contains little record of large dataset statistical sampling, but it is urgently needed to demonstrate the suitability of different vortex models and to clarify the relationships between them. As stated in the introduction, the large-scale streamwise vortex model provides a good explanation of the coherent structures of Q2/Q4 events and the spanwise distances between adjacent streaky structures near the water surface (which is about 2*H*). However, the large-scale streamwise vortex model is relatively coarse and represents a large structure (Figure 5), and it cannot accurately explain the continuous development of streaks from the inner region to the outer region. The attached eddy vortex model cannot provide a precise organized structure for the large vortices that accumulate in turbulence. The hairpin vortex model requires more usage and analysis to show its suitability.

Vortex models are limited. However, our research into the characteristic dimensions of streaky structures across the entire water depth, described in this study, leads us to conclude: the streamwise vortex model, the attached eddy vortex model, and the hairpin vortex model are all suitable models in certain circumstances.

We used quantitative analysis to develop a theoretical model in which packets of attached eddy vortices self-organize and accumulate along the flow direction, thereby forming a cumulative vortex structure, the streamwise vortex. Figure 16 shows that many attached eddy vortices are connected along the flow direction to form the large-scale structure of a streamwise vortex. This behavior provides more details about the formation of large-scale streamwise vortexes. The model explains the characteristics of streamwise vortices (Q2/Q4 events and low- and high-speed streaky structures) and linear streaks based on water depth. Our investigation into the spatial relationships between spanwise vortex density and streaky structures shows that the legs of the hairpin vortex model envelop low-speed streaks. These low-speed hairpin vortex legs can be organized and accumulated into larger-scale quasi-streamwise vortices (Figure 16).

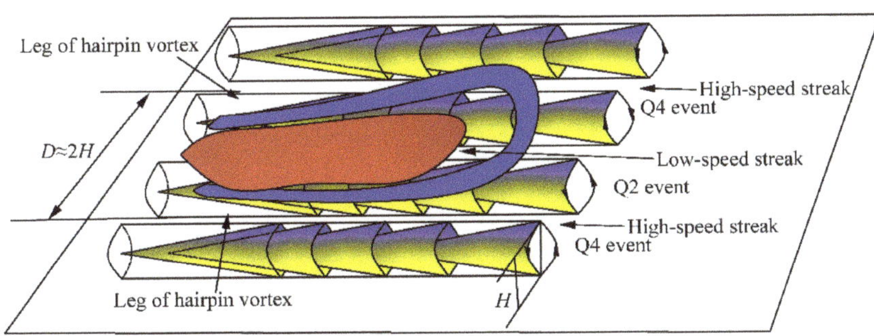

Figure 16. Theoretical models of a streamwise vortex, an attached eddy vortex, and a hairpin vortex in the *x*–*z* plane.

Further analysis of the details of the vortex models led us to propose a simple hypothesis: the three coherent structures, modeled individually as a streamwise vortex, while an attached eddy vortex and a hairpin vortex both exist separately in turbulent flow. It is likely that they are all manifestations of the same turbulent structure under different paradigmatic approaches, as shown in Figure 16.

4.2. Conclusions

We based this study on 30 sets of closed-channel flow DNS data. Image processing was employed to identify low- and high-speed streaks, using a detection function and a threshold value, C_t. Statistical methods were used to calculate the characteristic dimensions of both low- and high-speed streaks. We investigated three models of coherent structures (streamwise vortex, attached eddy vortex, and hairpin vortices) and demonstrated their application. Analysis of the characteristic dimensions of streaky structures and vortices and further analysis of the relationships among the three vortex models led us to suggest a straightforward hypothesis. The results we obtained are summarized as follows.

(1) The average width of streaks and the average distance between adjacent streaks that we observed are consistent with the results of previous studies, which indicates the suitability of our method of identifying and calculating both low- and high-speed streaky structures.
(2) The development of streaks from the inner turbulent region to the outer region is a continuous process. The length of streaky structures increases linearly with the water depth, and it is approximately twice the water depth. This result also shows the suitability of both the streamwise vortex and the funnel vortex models.
(3) The spanwise vortex density in the x–z plane is greatest within low-speed streaks, intermediate in the in-between region, and least in the high-speed streaks. We infer that the legs of the hairpin vortices envelop the low-speed streaky structures to move in the streamwise direction and conclude that the hairpin vortex model provides a suitable representation.
(4) The theoretical model of the locations in the x–z plane of streamwise vortices, attached-eddy vortices and hairpin vortices established the possibility of the coexistence of three vortex structures; this recognition increases our understanding of the mechanics of coherent structures in turbulent flows.

Author Contributions: H.W. performed the data collection, data processing, and manuscript preparation; G.P. assisted with the data analysis. All authors participated in the revision of the manuscript and graphics.

Funding: The study is financially supported by the National Natural Science Foundation of China (Grant No. 51709047), the CRSRI Open Research Program (Grant No. CKWV2018461/KY), and the Open Research Fund of State Key Laboratory of Simulation and Regulation of Water Cycle in River Basin (China institute of Water Resources and Hydropower Research, Grant No. IWHR-SKL-KF201815).

Conflicts of Interest: The authors declare no conflicts of interest.

Abbreviations

Parameter	Description Unit
A_s	Total area of low- or high-speed streaky structures
A_t	Total area of the flow field (2048 × 1536)
$Ct_{(y+)}$	Streak threshold at y^+
$d_{(r)}$	Dimensionless spanwise distance at each y_+
D	Nondimensional spanwise distance
$\overline{F_d}$	Detection function value of average vortex velocity
F_d	Dimensionless value of detection function
m	Total number of rows of the flow field
N_x	Grid numbers in the x direction
N_y	Grid numbers in the y direction
N_z	Grid numbers in the x direction
N_{vortex}	Number of spanwise vortices at position y^+
P_s	Area percentages of low- and high-speed streaky structures
u	Instantaneous velocity in the x direction m/s
u_*	Friction velocity m/s

uij	Velocity at the position of vortex core
$\bar{u}$	Average velocity of spanwise vortices in the x–z plane m/s
$\bar{u}'$	Fluctuating velocity m/s
u'	Streamwise velocity fluctuation m/s
v	Instantaneous velocity in the y direction m/s
w	Instantaneous velocity in the z direction m/s
W	Average nondimensional width
x	Streamwise direction
Δx	Grid resolution in the x directions
y^*	Inner scale
y	Vertical direction
Δyc	Wall-normal grid spacing at the channel left
z	Spanwise direction
Δz	Grid resolution in the z directions
(m, n)	Grid position in the x–z plane
$ustd(y^+)$	Standard deviation of the streamwise velocity at y^+
λci	Two-dimensional swirling-strength 1/s
Λci	Dimensionless swirling strength
ω_z	Fluctuating spanwise vorticity 1/s
$\Lambda_{ci}^{rms}(y)$	Local root mean square of Λci at the wall-normal position y
Ωci	Normalized swirling strength
Π^+	Vortex population density
υ	Kinematic viscosity cm^2/s

References

1. Zhong, Q.; Chen, Q.G.; Wang, H.; Li, D.X.; Wang, X.K. Statistical analysis of turbulent super-streamwise vortices based on observations of streaky structures near the free surface in the smooth open channel flow. *Water Resour. Res.* **2016**, *52*, 3563–3578. [CrossRef]
2. Kline, S.J.; Reynolds, W.C.; Schraub, F.A.; Runstadler, P.W. The structure of turbulent boundary layers. *J. Fluid Mech.* **1967**, *30*, 741–773. [CrossRef]
3. Kähler, C.J. Investigation of the spatio-temporal flow structure in the buffer region of a turbulent boundary layer by means of multiplane stereo PIV. *Exp. Fluids* **2004**, *36*, 114–130. [CrossRef]
4. Carlier, J.; Stanislas, M. Experimental study of eddy structures in a turbulent boundary layer using particle image velocimetry. *J. Fluid Mech.* **2005**, *535*, 143–188. [CrossRef]
5. Lagraa, B.; Labraga, L.; Mazouz, A. Characterization of low-speed streaks in the near-wall region of a turbulent boundary layer. *Eur. J. Mech. B Fluids* **2004**, *23*, 587–599. [CrossRef]
6. Lin, J.; Laval, J.; Foucaut, J.; Stanislas, M. Quantitative characterization of coherent structures in the buffer layer of near-wall turbulence. *Exp. Fluids* **2008**, *45*, 999–1013. [CrossRef]
7. Nakagawa, H.; Nezu, I. Structure of space-time correlations of bursting phenomena in an open-channel flow. *J. Fluid Mech.* **1981**, *104*, 1–43. [CrossRef]
8. Shvidchenko, A.B.; Pender, G. Macroturbulent structure of open-channel flow over gravel beds. *Water Resour. Res.* **2001**, *37*, 709–719. [CrossRef]
9. Roy, A.G.; Buffin-Belanger, T.; Lamarre, H.; Kirkbride, A.D. Size, shape and dynamics of large-scale turbulent flow structures in a gravel-bed river. *J. Fluid Mech.* **2004**, *500*, 1–27. [CrossRef]
10. Wang, H.; Zhong, Q.; Wang, X.K.; Li, D.X. Quantitative characterization of streaky structures in open-channel flows. *J. Hydraul. Eng.* **2017**, *143*, 04017040. [CrossRef]
11. Sukhodolov, A.N.; Nikora, V.I.; Katolikov, V.M. Flow dynamics in alluvial channels: The legacy of Kirill V. Grishanin. *J. Hydraul. Res.* **2011**, *49*, 285–292. [CrossRef]
12. Fox, J.F.; Patrick, A. Large-scale eddies measured with large scale particle image velocimetry. *Flow Meas. Instrum.* **2008**, *19*, 283–291. [CrossRef]
13. Panton, R.L. Overview of the self-sustaining mechanisms of wall turbulence. *Prog. Aerosp. Sci.* **2001**, *37*, 341–383. [CrossRef]

14. Gulliver, J.S.; Halverson, M.J. Measurements of large streamwise vortices in an open-channel flow. *Water Resour. Res.* **1987**, *23*, 115–123. [CrossRef]
15. Imamoto, H.; Ishigaki, T. Visualization of longitudinal eddies in an open channel flow. In Proceedings of the Fourth International Symposium on Flow Visualization IV, Paris, France, 26–29 August 1986; pp. 323–337.
16. Rashidi, M.; Banerjee, S. Turbulence structure in free-surface channel flows. *Phys. Fluids* **1988**, *31*, 2491–2503. [CrossRef]
17. Townsend, A.A. Equilibrium layers and wall turbulence. *J. Fluid Mech.* **1961**, *11*, 97–120. [CrossRef]
18. Adrian, R.J.; Marusic, I. Coherent structures in flow over hydraulic engineering surfaces. *J. Hydraul. Res.* **2012**, *50*, 451–464. [CrossRef]
19. Nezu, I.; Rodi, W. Open-channel flow measurements with a laser Doppler anemometer. *J. Hydraul. Eng.* **1986**, *112*, 335–355. [CrossRef]
20. Nezu, I.; Nakagawa, H.; Jirka, G.H. Turbulence in open-channel flows. *J. Hydraul. Eng.* **1994**, *120*, 1235–1237. [CrossRef]
21. Del Alamo, J.C.; Jiménez, J.; Zandonade, P.; Moser, R.D. Scaling of the energy spectra of turbulent channels. *J. Fluid Mech.* **2004**, *500*, 135–144. [CrossRef]
22. Herpin, S.; Stanislas, M.; Soria, J. The organization of near- wall turbulence: A comparison between boundary layer SPIV data and channel flow DNS data. *J. Turbul.* **2010**, *11*, 1–30. [CrossRef]
23. Kaftori, D.; Hetsroni, G.; Banerjee, S. Particle behavior in the turbulent boundary layer. I. Motion, deposition, and entrainment. *Phys. Fluids* **1995**, *7*, 1095–1106. [CrossRef]
24. Adrian, R.J. Hairpin vortex organization in wall turbulence. *Phys. Fluids* **2007**, *19*, 041301. [CrossRef]
25. Zhou, J.; Adrian, R.J.; Balachandar, S.S.; Kendall, T.M. Mechanisms for generating coherent packets of hairpin vortices in channel flow. *J. Fluid Mech.* **1999**, *387*, 353–396. [CrossRef]
26. Wu, Y.; Christensen, K.T. Population trends of spanwise vortices in wall turbulence. *J. Fluid Mech.* **2006**, *568*, 55–76. [CrossRef]
27. Chen, H.; Adrian, R.J.; Zhong, Q.; Wang, X.K. Analytic solutions for three dimensional swirling strength in compressible and incompressible flows. *Phys. Fluids* **2014**, *26*, 81701. [CrossRef]
28. Zhong, Q.; Li, D.X.; Chen, Q.G.; Wang, X.K. Coherent structures and their interactions in smooth open channel flows. *Environ. Fluid Mech.* **2015**, *15*, 653–672. [CrossRef]

Article

Viscosity Controls Rapid Infiltration and Drainage, Not the Macropores

Peter Germann

Faculty of Science, Institute of Geography, University of Bern, 3012 Bern, Switzerland; pf.germann@bluewin.ch

Received: 12 November 2019; Accepted: 20 January 2020; Published: 24 January 2020

Abstract: The paper argues that universal approaches to infiltration and drainage in permeable media pivoting around capillarity and leading to dual porosity, non-equilibrium, or preferential flow need to be replaced by a dual process approach. One process has to account for relatively fast infiltration and drainage based on Newton's viscous shear flow, while the other one draws from capillarity and is responsible for storage and relatively slow redistribution of soil water. Already in the second half of the 19th Century were two separate processes postulated, however, Buckingham's and Richards' apparent universal capillarity-based approaches to the flow and storage of water in soils dominated. The paper introduces the basics of Newton's shear flow in permeable media. It then presents experimental applications, and explores the relationships of Newton's shear flow with Darcy's law, Forchheimer's and Richards' equations, and finally extends to the transport of solutes and particles.

Keywords: wetting shock fronts; shear flow; viscosity; capillarity; kinematic waves

---

1. Introduction

Infiltration is the transgression of liquid water from above the surface of the permeable lithosphere to its interior, while drainage refers to liquid water leaving some of its bulk. For example, within about three weeks after liquid manure applications, bad odors appeared in drinking water wells at depths between 10 to 50 m in English Chalk [1]. In the same geological formations, the annual surplus of the water balance moved down with about 1 m year^{-1} as profiles of isotope ratios of $^{18}O/^{16}O$ revealed [2]. The contrasting observations lead to ratios between 150 and 700 of the velocities of the 'odor' front vs. the isotope front that illustrate well the difference between rapid and slow infiltration.

Fast flow and transport are usually attributed to preferential flow in the macropore domain, while slower movements are supposedly due to flow in the capillary domain as expressed with Richards' [3], allowing for exchanges between the two domains as, for instance [4] and [5] have recently compiled. This contribution favors a dual-process approach to infiltration and drainage over dual-porosity approaches, thus avoiding the a-priori delineation between 'macropores' and the remaining porosity.

Beginning in the middle of nineteenth century, the next section summarizes the evolution of permeable media flow concepts for both, saturated and unsaturated flows. The third section provides the base for Newton's shear flow, stressing vertical viscous flow during rapid infiltration, while capillarity is assigned to the much slower redistribution of soil moisture but in all directions. The fourth section provides experimental evidence, and is followed by the conclusions.

2. Evolution of Infiltration and Drainage Concepts

This section provides some milestones on the way to a dual-process approach to infiltration and drainage, where Newton's shear flow covers fast and gravity driven flows, while Richards' capillary flow is considered to deal mainly with slower capillary rises and redistributions. By no means is this section intended to cover the history of infiltration and drainage.

2.1. Steady Saturated Flow

In the mid-19th century, there was an increasing interest in flows in saturated soils and similarly permeable media. Hagen, a German hydraulic engineer, and Poiseuille (1846) [6] a French physiologist, independently analyzed laminar flow in thin capillary tubes. Darcy (1856) [7], in the quest of designing a technical filtration system for the city of Dijon, empirically developed the concept of hydraulic conductivity as proportionality factor of flow's linear dependence on the pressure gradient. Dupuit (1863) [8] expanded Darcy's law to two dimensions as perpendicular and radial flow between two parallel drainage ditches and towards a groundwater well, respectively. Forchheimer (1901) [9] added an expression to Darcy's law that accounts for high pressures, steep gradients, and high flow velocities in the liquid.

2.2. Early Studies on Flow in Unsaturated Soils

Schumacher (1864) [10], a German agronomist, was probably the first who considered capillarity as the cause for simultaneous flows of water and air in partially water-saturated soils. He qualitatively compared the rise of wetting fronts in soil columns with the rise of water in capillary-sized glass tubes, and concluded that the wetting fronts rise higher but slower in finer textured soils compared with coarser materials. He also infiltrated water in columns of undisturbed soil and found that infiltration fronts progressed much faster than the rising wetting fronts. He suggested two separate processes for the two flow types: (i) slower capillary rise and (ii) faster infiltration, however, without further dwelling on the latter. Lawes et al. (1882) [11] concluded from the chemical composition of the drain from large lysimeters at the Rothamsted Research Station that "The drainage water of a soil may thus be of two kinds (1) of rainwater that has passed with but little change in composition down the open channels of the soil; or (2) of the water discharged from the pores of a saturated soil." [11] prioritized two separate flow paths to explain the observations.

2.3. Universal Capillary Flow in Unsaturated Soils

During the second half of the 19th century, modern irrigation agriculture spread in semi-arid areas, thus increasing the demand for better understanding the soil-water regime. Buckingham (1907) [12], working on a universal approach to the simultaneous storage and flow of water and air in soils, postulated the relationship between the capillary potential ψ Pa and the volumetric water content θ m^3 m^{-3}, also known as the water retention function, retention curve, or water release curve. The capillary potential follows from the Young–Laplace (1805) [13] relationship, stating that the pressure difference between a liquid and the adjacent gas phase increases inversely proportional to the radius of the interface. Capillary potential emerges as energy per unit volume of water in the permeable medium due to the water's surface tension. Canceling energy and volume with one length leads to force per area. Because the menisci's surfaces are bent into the liquid, $\psi < 0$, where $\psi = 0$ corresponds to the air pressure as reference. In addition to the specific weight of the soil water, [12] introduced the spatial gradient of ψ as the other major driving force. Besides infiltration, this stroke of a genius accounts for the redistribution of soil water in all directions, evaporation across the soil surface, transpiration via roots, and capillary rise from perched water including groundwater tables. In analogy to Fourier's (1822) [14] and Ohm's (1825) [15] laws for heat flow and electrical current, and Darcy's law for water flow in saturated porous media, [12] postulated the hydraulic conductivity for flow in unsaturated porous media as function of either $K(\theta)$ or $K(\psi)$ m s^{-1}. According to Or (2018) [16], the British meteorologist Richardson (1922) [17] was most likely the first who introduced a diffusion type of K-ψ-θ-relationship in the quest of quantifying water exchange between the atmosphere and the soil as lower boundary of the meteorological system. A second-order partial differential expression became necessary because ψ depends on θ, and both their temporal variations on flow, while flow itself is driven by the gradient of ψ.

The race was on to the experimental determination of the K-ψ-θ-relationships. For instance, Gardner et al. (1922) [18] used plates and blocks of fired clay with water-saturated pores fine enough to hydraulically connect the capillary-bound water within soil samples with systems outside them. [3] applied the technique to the construction of tensiometers that directly measure ψ within an approximate range of $0 > \psi >\approx -80$ kPa. With the pressure plate apparatus he measured ψ-θ-relationships and determined hydraulic conductivity $K(\psi$ or $\theta)$. Similar to [17], he presented a diffusion-type approach to the transient water flow in unsaturated soils. Numerous analytical procedures evolved for solving the Richards equation, among them a prominent series of papers by J.R. Philip [19]. Moreover, van Genuchten (1980) [20] developed mathematically closed forms of K-ψ-θ-relationships that provide the base for the many hues of HYDRUS i.e., various numerical simulation packages dealing with flow and storage of water and solutes in unsaturated soils (Simunek et al., 1999, [21]).

2.4. Early Alternatives to Universal Capillary Flow

In his quest of demonstrating the benefit of forests and reforestations on controlling floods and debris flows from steep catchments in the Swiss Alps and Pre-Alps, Burger (1922) [22] measured in situ the time lapses Δt_{100} for the infiltration of 100 mm of water into soil columns of the same length. In the laboratory, he determined the air capacity AC m^3 m^{-3} of undisturbed samples collected near the infiltration measurements, where AC is the difference of the specific water volume after standardized drainage on a gravel bed and complete saturation.

Veihmeyer (1927) [23] investigated water storage in soils with the aim of scheduling irrigation. He proposed the water contents at field capacity (FC) and at the permanent wilting point (PWP) as upper and lower thresholds of plant-available soil water. Free drainage establishes FC a couple of days after a soil was saturated and evapo-transpiration was prevented. The water loss before achieving FC is referred to as 'drainable' or 'gravitational' soil water. PWP is considered θ at $\psi = -150$ kPa ($= -15$ bar).

2.5. Dual Porosity Approaches

It became unavoidable that concepts based on Buckingham's (1907) [12] fundamental and seminal work contradicted with practical and field-oriented research. Veihmeyer (1954) [24], for instance, stated "Since the distinction between capillary and other 'kinds' of water in soils cannot be made with exactness, obviously a term such as non-capillary porosity cannot be defined precisely since by definition it is determined by the amount of 'capillary' water in the soils". Additionally, progress in field instrumentation as well as in computing techniques allowed for producing and processing large data sets including the numerical solution of Richards equation. In the late 1970s, the development increasingly unveiled substantial discrepancies between measurements and the numerous approaches to water movement in unsaturated soils based on [3] capillarity-dominated theory. Particularly disturbing were observations on wetting fronts advancing much faster than expected from the Richards approach.

The mid-1970s initiated the thread of dual-porosity approaches to preferential flows. Bouma et al. (1977) [25] were among the first to introduce the term macropores in view of non-equilibrium in ψ-θ-relationships, while the compilation by Beven and Germann (1982) [26] on the subject is still referred to today. The discussion has gradually moved from macropore flow to preferential flow that summarizes all the flows in unsaturated porous media not obeying the Richards equation [4]. Numerous reviews on macropore flows, preferential flows, non-equilibrium flows, and non-uniform flows in permeable media appear periodically, and only the latest are here referenced [5,27].

Beven (2018) [28] argued that, for about a century, the hardly questioned preference given to capillarity denied recognition of concepts considering flow along macropores, pipes, and cracks. Indeed, there is an increasing number of contributions focusing on the dimensions and shapes of flow paths, their 3-D imaging, and trials to derive flows from them [4]. However, there is hardly an approach capable of applying the wealth of information about the paths to the quantification of flow.

Ignoring Veihmeyer's (1954) [23] warning, the attraction of research on flow paths is so dominant that, for instance, Jarvis et al. (2016) [4] flatly denied the applicability of Hagen–Poiseuille concepts to flow in soils. Moreover, advanced techniques of infiltration with non-Newtonian fluids led so far just to the description of path structures rather than more directly to the flow process [29]. Wide-spread research in the types, dimensions, and shapes of 'macropores' and their apparent relationships to flow and transport mostly pivot around Richards equation that is numerically applied to either macropore-/micropore-domains or by modelling flow and solute transport in the macropore domain with separate rules yet still maintaining a Richards-type approach to micropore-flow. Both types of approaches allow for exchanges of flow and solutes between the two domains. Imaging procedures visualize flow in 2-D and 3-D in voids as narrow as some 10 μm, rising hope that the wealth of information gained so far at the hydro-dynamic scale will eventually lead to macroscopic models at the soil profile scale of meters [4]. Thus, Beven's (2018) [28] denial of progress in infiltration research is here carried a step further. The obsession with pores, channels, flow paths, and their connectivity, tortuosity, and necks actually retarded research progress towards more general infiltration that has to be based on hydro-mechanical principles.

2.6. *Early Search for Alternatives*

The alternative approach should be based on the same principles as Hagen–Poiseuille's and Darcy's laws, whereas Newton's (1729) [30] shear flow (Nsf) appears as a solid and suitable foundation. The approach then should close the gap of one to two orders of magnitude of hydraulic conductivity between saturated flow and flow close to saturation [31].

Unearthing Burger's (1922) [22] data, [32] found an encouraging coefficient of determination of $r^2 = 0.77$ when correlating via a Hagen–Poiseuille approach 76 pairs of Δt_{100}- and AC-values. Beven and Germann (1981) [33] modelled laminar flow in tubes and planar cracks, and proposed kinematic wave theory according to Lighthill and Whitham (1955) [34] as analytical tool for handling Nsf. Germann (1985) [35] applied the theory successfully to data from an infiltration-drainage experiment carried out on a block of polyester consolidated coarse sand that experimentally support the flow-(q-) version of Nsf, where q m s^{-1} is volume flux density.

Dye experiments confirmed expectations, not the least by purposefully setting the initial and boundary conditions. [36], for instance, demonstrated preferential flow along earthworm burrows down to the 0.4 m depth as well as radially away from the channels by inundating with rhodamine dye and a bromide solution the tops of columns of undisturbed soil. Following [11], who reported fast drainage, Germann (1986) [37] assessed the arrival times of precipitation fronts in the Coshocton lysimeters. Accordingly, rains of 10 mm d^{-1} sufficed to initiate or increase drainage flow within 24 h at the 2.4 m depth when θ in the upper 1.0 m of the soil was at or above 0.3 m^3 m^{-3}. The observations result in wetting front velocities greater than 3×10^{-5} m s^{-1}.

2.7. *Development of Newton's Shear Flow Approach to Infiltration and Drainage*

The advent of TDR- and data recording techniques fostered the water-content-(w-) version of Nsf (where w m^3 m^{-3} refers to the mobile water content) applicable to in-situ infiltration experiments using controlled sprinkler irrigation and producing time series of θ at various depths in soil profiles. Discussions with L. Dipietro and V. P. Singh linked experiments with basic concepts [38]. Investigations of acoustic velocities across a column of an undisturbed soil during infiltrations, [39] proved independently from the capillarity narrative that rapidly infiltrating water is under atmospheric pressure. Originally, macropore flow was the prominent research topic. Shortfalls of the capillarity-dominated concepts emerged when juxtaposed to matured Nsf-versions [40]. Hence, this paper is the première of consequently carrying through the dual-process approach to infiltration and drainage.

There were projects leading off the track. [41], for instance, proposed a kinematic wave approach to infiltration that included a sink function accounting for water abstraction from the moving to the sessile parts. Later rigorous testing of the approach against real data unveiled the misconception. Additionally, [42] superimposed a bundle of rivulets to model data. That again turned out as impracticable. [43] were looking for FC in soil profiles after the cessation of infiltration. However, repeated experiments showed quite variations in expected constant final FCs.

2.8. State of Nsf

Gradually, some of the subordinate concepts lined out above get referenced. However, as of yet, outside the author's immediate environment of research, there is hardly an elaborated application of Nsf to infiltration and drainage. Unavoidably, this leads to the impression of self-indulgence, particularly in view of the paper's lopsided list of references. The Guest Editor's invitation to its submission may ease the charge. Moreover, although [26] mainly outlined the need for research on 'macropore flow' and did not offer much of research guidance, that paper is still frequently referenced, indicating that researchers have hardly agreed on a basic approach to the problem as they did and still do, for instance, on the Richards equation.

3. Newton's Shear Flow in Permeable Media

The section summarizes the basics of Newton's shear flow (Nsf) in permeable media. Detailed derivations of the relationships have previously been published [44–46], thus only the major expressions are here presented. The approach is laid out at the hydro-mechanical scale of spatio-temporal process integration, allowing for its easy handling with analytical expressions, yet under strict observance of the balances of energy, momentum, and mass (i.e., the continuity requirements).

The interior of a permeable solid medium contains connected flow paths that are wide enough to let liquids pass across its considered volume. The definition purposefully avoids further specification of the flow paths' shapes and dimensions. Water supply to the surface is thought of a pulse $P(q_S,T_B,T_E)$, where q_S m s^{-1} is constant volume flux density from the pulse's beginning at T_B to its ending at T_E, both s. (The subscript S refers to the surface of the permeable medium). The pulse initiates a water content wave (*WCW*) of mobile water that is conceptualized as a film gliding down the paths of a permeable medium according to the rules of Newton's shear flow. The *WCW* is the basic unit of shear flow whose hydro-mechanical properties are going to be presented below.

Referring to Figure 1, the parameters film thickness F m and specific contact length L m m^{-2} per unit cross-sectional area A m^2 of the medium specify a *WCW*. Regardless of the thickness of F, atmospheric pressure prevails within the film. A *WCW* supposedly runs along the flow paths while forming a discontinuous and sharp wetting shock front at $z_W(t)$. The *WCW* partially fills the voids of the upper part of the medium within $0 \le z \le z_W(t)$ with the mobile water content $w(z,t)$ m^3 m^{-3}, where $w \le \varepsilon - \theta_{ante}$, where ε and θ_{ante} are porosity and antecedent θ, respectively, both m^3 m^{-3}. The lower part $z > z_W(t)$ remains at θ_{ante}. The coordinate z m originates at the surface and points positively down. The water film of the *WCW* consists of an assembly of parallel laminae, each df m thick and moving with the celerity of $c(f)$ m s^{-1}. Celerity refers to the wave velocity, for instance of a lamina. Newton (1729) [30] defined viscosity as "The resistance, arising from the want of lubricity in the parts of a fluid, is, caeteris paribus, proportional to the velocity with which the parts of the fluid are separated from each other." In our case, the definition translates to the shear stress $\varphi(f)$ Pa in the unit area of $L \times A \times z_W(t)$ m^2 per unit volume $A \times z_w(t)$ m^3 of the medium at distance $0 \le f \le F$ m from the soil-water interface (*SWI*), acting in the direction opposite to gravity. $\varphi(f)$ balances the weight of the film from f to F. Integration of the shear stress balance from 0 to f under consideration of the non-slip condition of $c(0) = 0$ leads to the parabolic celerity profile of the laminae within the film. The differential flow of a lamina is $dq(f) = L \times c(f) \times d(f)$. Its integration from 0 to F i.e., from the *SWI* to the air-water interface (*AWI*), leads to the volume flux density of the entire film as:

$$q(F,L) = \frac{g}{3 \cdot \eta} \cdot L \cdot F^3 \tag{1}$$

m s^{-1}, where g (=9.81 m s^{-2}) is acceleration due to gravity and η ($\approx 10^{-6}$ m^2 s^{-1}) is temperature-dependent kinematic viscosity. The volume of mobile water per unit volume of the permeable medium from the surface to $z_W(t)$ amounts to:

$$w(F, L) = F \cdot L \tag{2}$$

m^3 m^{-3}. The constant velocity of the wetting shock front follows from the volume balance as:

$$v_W(F) = \frac{q(F, L)}{w(F, L)} = \frac{g}{3 \cdot \eta} \cdot F^2 \tag{3}$$

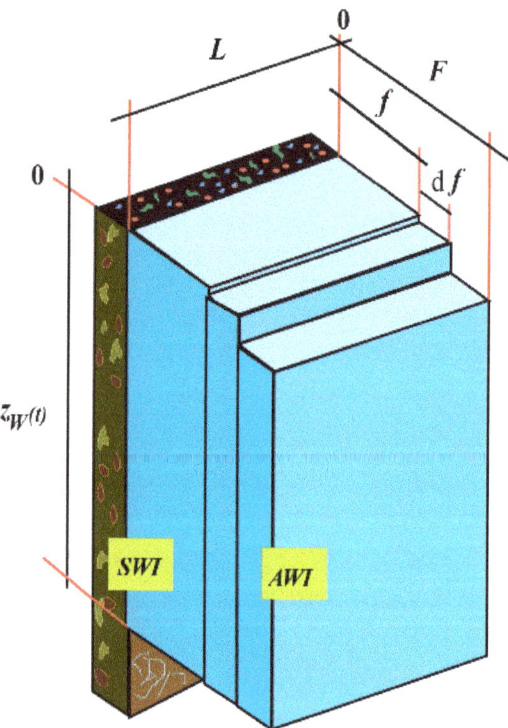

Figure 1. Film flow along a vertical plane, where F m is the film thickness, f m is the thickness variable, df m is the thickness of a lamina, $z_W(t)$ m is the vertical position of the wetting shock front as function of time t s, and L m m^{-2} is the specific contact length between the moving water film and the sessile parts of the permeable medium per its unit horizontal cross-sectional area A m^2. *AWI* and *SWI* are the air-water and the solid-water interfaces, respectively ([44]; with permission from the publisher).

Thus, the position of the wetting shock front as function of time becomes:

$$z_W(t) = v_W \cdot (t - T_B) = (t - T_B) \cdot \frac{g}{3 \cdot \eta} \cdot F^2 \tag{4}$$

m s^{-1}. Accordingly, L m m^{-2} is not only the specific contact length of the film per A but, under consideration of $z_W(t)$, L m^2 m^{-3} evolves as the specific vertical contact area of the *WCW* with the sessile parts of the permeable medium per unit volume. It thus turns into the locus of exchange of momentum, heat, capillary potential, water, solutes, and particles between the *WCW* and the sessile parts of the medium.

Equations (1)–(4) hold during infiltration i.e., $T_B \leq t \leq T_E$. Input ends abruptly at T_E and at $z = 0$ i.e., $q_S \rightarrow 0$, when and where the *WCW* collapses from $f = F$ to $f = 0$. All the rear ends of the laminae are released at once at $z = 0$. The outermost lamina moves the fastest with the celerity of the draining front as:

$$c_D = \frac{dq(F)}{dw} = \frac{dq(F)}{L \cdot df} = \frac{g}{\eta} \cdot F^2 = 3 \cdot v_W \tag{5}$$

m s^{-1}. While v_W follows from the volume balance Equation (3), the three times faster c_D follows from the parabolic function *c*(*f*) (see, for instance, Germann and Karlen, 2016). The outer most lamina moves with c_D also during $0 \leq t \leq T_E$. Therefore, the slower moving wetting shock front continuously intercepts the faster moving laminae. This results in a sharp wetting shock front at $z_W(t)$ with θ_{ante} + w between the surface and the wetting shock front and θ_{ante} ahead of it. The wetting shock front presents a discontinuity of the *WCW*. The rear end of the draining front that was released at T_E catches up with wetting shock front at time T_I s that follows from setting $v_W \times (T_I - T_B) = c_D \times (T_I - T_E)$ as

$$T_I = \frac{1}{2} \cdot (3 \cdot T_E - T_B) \tag{6}$$

Thus, T_I is exclusively an expression of the pulse duration. The wetting shock front intercepts the draining front at depth:

$$Z_I = (T_E - T_B) \cdot F^2 \cdot \frac{g}{2 \cdot \eta} \tag{7}$$

m. The two expressions T_I and Z_I represent the spatio-temporal scale of a *WCW*. The rear ends of all the other laminae move with decreased celebrities, ultimately leading to the spatio-temporal distribution of the mobile water content from the surface to the wetting shock front as:

$$w(z,t) = L \cdot \left(\frac{\eta}{g}\right)^{1/2} \cdot z^{1/2} \cdot (t - T_E)^{-1/2} \tag{8}$$

After T_I and beyond Z_I the draining front disappears and $v_W(z,t)$ decreases with time and depth, leading to the position of the wetting shock front as:

$$z_W(t) = \left(\frac{3 \cdot V_{WCW}}{2 \cdot L}\right)^{2/3} \cdot \left(\frac{g}{\eta}\right)^{1/3} \cdot (t - T_E)^{1/3} \tag{9}$$

with the mobile water content of:

$$w(t)\big|_{z_W} = \left(\frac{\eta}{g}\right)^{1/3} \cdot \left(\frac{3 \cdot V_{WCW}}{2}\right)^{1/3} \cdot (t - T_E)^{-1/3} \cdot L^{2/3} \tag{10}$$

while the volume flux density at the wetting shock front becomes:

$$q(t)\big|_{z_W} = \frac{V_{WCW}}{2 \cdot (t - T_E)} \tag{11}$$

where $V_{WCW} = q_S \times (T_E - T_B)$ m is the total volume of the *WCW*. Figure 2 illustrates the relationships describing a *WCW*.

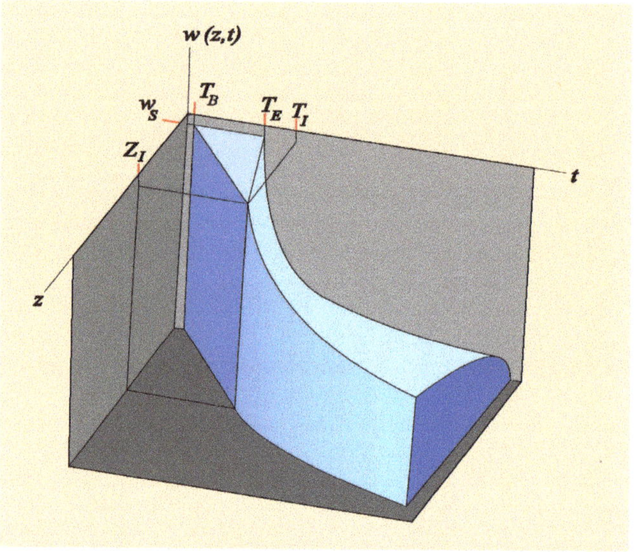

Figure 2. Schematic representation of a water-content wave *WCW*, where the $w(z,t)$-axis represents the mobile water content, t and z are the axes of time and depth; w_S indicates the mobile water content that follows from q_S; T_B and T_E s are the beginning and ending times of the water pulse $P(q_S, T_B, T_E)$ that hits the surface at $z = 0$; T_I and Z_I indicate time and depth of the wetting front intercepting the draining front. The wetting shock front continues beyond $t > T_I$ and $z > Z_I$ as curve along time and depth. ([44]; with permission from the publisher).

A *WCW* i.e., $w(z,t)$,translates easily into a volume flux density wave according to:

$$q(z,t) = b \cdot w(z,t)^3 \tag{12}$$

with $b = g/\left(3 \cdot L^2 \cdot \eta\right)$ m s^{-1}. The approach applies to laminar flow as assessed with the Reynolds number:

$$Re = \frac{F \cdot v}{\eta} = \frac{F^3 \cdot g}{3 \cdot \eta^2} = \left(\frac{3 \cdot v^3}{g \cdot \eta}\right)^{1/2} \tag{13}$$

$Re \leq 1$ strictly defines laminar flow; however, depending on the application, $Re > 1$ might be tolerable, yet within an undisclosed range.

The parameters film thickness F m and specific contact area L m^2 m^{-3} of a *WCW* due to a pulse $P(q_S,T_B,T_E)$ follow from time series of either volumetric water contents $\theta(Z,t)$ or from volume flux density $q(Z,t)$, where Z (m) represents a specific depth in the permeable medium. Presumably, F and L get established shortly after T_B. Their sizes are due to q_S and the actual properties of the permeable medium, such as θ_{ante} and related flow paths supporting F and L. Predictions of F and L are currently undisclosed, thus restricting the approach to *a posteriori* data interpretation.

4. Examples

The section provides examples in support and as illustrations of Nsfa. More, and more elaborated examples are presented in [44].

4.1. Experimental Support of Newton's Shear Flow in Permeable Media

Constant wetting front velocities, Equation (3): Already [47] observed constant velocities of wetting fronts during finger flow in sandboxes down to the 0.9-m depth. [48] reported the same kind of observations over a depth range of 2 m during infiltration in the Kiel Sand Tank, while [49] found constant wetting front velocities under natural precipitation regimes across the upper 1.5 m of the undisturbed soil in a large weighing lysimeter. With 34 FTDR-probes in three boreholes, [50] followed infiltration in an ancient sand dune in Israel. From their data, [44] deduced two constant wetting front velocities: one of about 5×10^{-6} m s^{-1} above the clay-rich layer at the 8-m depth, and one of about 2×10^{-6} m s^{-1} down to the 21 m depth. The two coefficients of determination (r^2) of the linear regressions of arrivals vs. depths of the wetting front amounted to 0.83 and 0.94, respectively. Constant wetting front velocities support the early stage of *WCWs*, Equation (3), and imply constant *F*.

Laminar flow: From more than 200 infiltration experiments, [51] reported a range of $10^{-5} < v_W < 2 \times 10^{-2}$ m s^{-1}. According to Equation (13), this leads to a range of $10^{-4} < Re < 1.6$. The upper limit is still close to laminar flow. In General, permissible upper limits of *Re* demand attention in view of the applicability of Nsf.

Atmospheric pressure behind wetting fronts: Atmospheric pressure in the *WCW* is a prerequisite of Nsf-theory. [48] antecedent capillary heads in the range from −1.2 to −0.8 m to collapse to about −0.2 m behind the wetting front. [42] found similar patterns of capillary-head collapses during infiltration into a column of a sandy-loam textured mollic Cambisol. Additionally, acoustic velocities during infiltration in a column of an undisturbed loess soil, [39] led [40] to the same conclusion of capillary head's break-up close to atmospheric pressure after the passing of the wetting front.

Cohesion of Newton's shear flow approach: The parameters *F* and *L* suffice to treat infiltration and drainage with Nsf, Equations (1) and (12). In principle, time series of either $\theta(Z,t)$ or $q(Z,t)$ permit calibration of the two parameters: [45] shows their calibration from $\theta(Z,t)$, while [47] provide the procedure from *q*-series. The latter authors and [44] also demonstrate the cohesion of the approach in that derivations from both time series resulted in the same amount of *L*.

4.2. Newton's Shear Flow vis-à-vis Darcy's Law and Forchheimer's Equation

Darcy's law mutates to the saturated-flow extension of unsaturated vertical shear flow under the following considerations. When replacing kinematic viscosity with the dynamic viscosity, $\mu = \rho \times \eta$, follows from Equation (1) that:

$$\theta_{\text{ante}} + w < \varepsilon \text{ and } \Delta p/\Delta z = \rho \times g \qquad q = \frac{F^3\,L}{3\,\mu}\,\rho\,g \tag{14}$$

where θ_{ante} m^{3}m^{-3} is the antecedent volumetric water content, ε m^{3}m^{-3} is porosity, $\Delta p/\Delta z$ Pa m^{-1} is the pressure gradient, ρ (= 1000 kg m^{-3}) is the density of water, and $\mu = \rho\,\eta$ Pa s is dynamic viscosity. At saturation we get:

$$\theta_{\text{ante}} + w = \varepsilon \text{ and } \Delta p/\Delta z = \rho \times g: \qquad q_{sat} = \frac{F_{sat}^3 L_{sat}}{3\mu}\rho g = K_{sat} \tag{15}$$

while an external pressure gradient leads to:

$$\theta_{\text{ante}} + w = \varepsilon \text{ and } \Delta p/\Delta z > \rho \times g: \qquad q(p) = \frac{F_{sat}^3 L_{sat}}{3\mu}\frac{\Delta p}{\Delta z} = q_{sat}\frac{\Delta p}{\Delta z \rho g} \tag{16}$$

Darcy's law states that *q* is proportional to $\Delta p/\Delta z$ i.e., volume flux density is a linear function of the flow-driving gradient with the proportionality factor K_{sat}. In view of the various dimensionalities of *w* proportional to (L^1, F^1), *v* proportional to (L^0, F^2), and *q* proportional to (L^1, F^3), linearity seems only possible if F_{sat} and L_{sat} remain constant and independent from *p* in the transition from only gravity-driven to pressure-driven shear flow at saturation i.e., in the transition from Equation (15) to Equation (16). The elaboration supports the linearity of Darcy's law, but it is not an independent proof of the law's linearity. As a consequence, $w = q/v$ also remains constant. Further, if $\theta_{\text{ante}} + w$

$= \varepsilon$, $dL_{sat}/dp = 0$ and $dF_{sat}/dp = 0$ then follows the hypothesis that $(F_{sat} \times L_{sat})$ represent $(F \times L)_{max}$ leading to K_{sat}. However, other combinations of $(F \times L)$ in unsaturated media are feasible that may lead to $q_{unsat} > q_{sat} = K_{sat}$. The presumption of $q_{unsat} > K_{sat}$ opens an new view on shear flow and the basics of infiltration that are in stark contrast to Richards' capillary flow, where a priori $K_{sat} > K(\theta$ or $\psi)$. Comparisons of the rates from field infiltration with K_{sat}-measurements in the laboratory vaguely support the presumption of [45] and [52] most recently provide experimental evidence of $q_{unsat} > K_{sat}$.

Atmospheric pressure prevails in a *WCW* in Newton's shear flow. However, parts of the the *WCW* may hit flow paths narrower than F, and $p > p_{atm}$ will occur. The difference $p > p_{atm}$ will dissipate almost instantaneously with the acoustic velocity in soil water between 300 and 800 m s^{-1} [53] if virtually no water has to be moved over short distances to the next path wider than F. However, increasing the distances between the paths that are wider than F leads to local pressure build up that may eventually lead to sessile water above a layer of reduced conductivity. Water saturation may occur from relatively short periods, for instance shortly after heavy rains, up to seasons as, for instance, mottles of chemically reduced and oxidized zones in soil profiles may illustrate. Drainage under $p > p_{atm}$ shows as a steeper recession than drainage under $p = p_{atm}$. [49] provide the details. [54] report $\theta(t)$-time series of TDR-measurements immediately above an impervious soil layer that deviates from those of free flow during infiltration and that indicates the evolution of sessile soil water.

Forchheimer (1901) [9] added a quadratic term to Darcy's law that accounts for momentum dissipation under high flow velocities that occur under high pressure gradients i.e.,

$$-\frac{dp}{dx} = \frac{\eta}{K}v + \frac{\rho}{k_2}v^2 \tag{17}$$

where K and k_2 represent hydraulic conductivity and turbulence, respectivly. Neither of the two conditions apply to Nsf in unsaturated permeable media in the lithosphere near its interface with the atmosphere. Moreover, [54] estimate the relative contribution of kinetic energy in the range of 10^{-6} to 10^{-4} in comparison with viscous momentum dissipation during flow in soils. Thus, Forchheimer's equation does not apply to Nsf.

4.3. Gravity and Capillarity

Gravity is the unique driving force in Newton's shear flow in unsaturated permeable media, while capillarity is responsible for the soil water's redistribution and vertical rise from saturated zones. [10] suggested a two-process approach to water flow and storage in partially saturated permeable media. While he recognized capillarity as responsible for the water's rise, and probably also its contribution to water redistribution in soil columns, he left open the mechanism behind infiltration. Here, the focus is on infiltration that is completely gravity-driven and viscosity-controlled, yet allowing for water abstraction due to capillarity from the mobile to the immobile part of the permeable system. Richards' universal equation for flow and storage in unsaturated porous media revolves around capillarity. It prominently, and unnecessarily, reduces the degrees of freedom of flow [39]. Unjustified, it relates the coefficient of momentum dissipation with antecedent flow conditions i.e., $K(\theta$ or $\psi)$, thus a priori excluding atmospheric pressure in the moving water. The exclusion leads to too slow advancements of wetting as Germann and Hensel (2006) [55] have demonstrated by comparing the results from HYDRUS-2 model [21] performances with observed infiltrations from more than 200 sites. Concentrating on gravity and viscosity liberates infiltration and drainage from the omnipresence of capillarity in soil hydrology with the benefit of avoiding the difficult definitions of non-equilibrium flow and the separation of macropores from the remaining pores. With respect to capillarity, the relative contribution of gravity to flow varies according to *cos* (α), where $\alpha°$ is the angle of deviation from the vertical. Thus, at *cos* (0°) = 1, as in the cases presented above, gravity's contribution is at maximum; it reduces to *cos* (90°) = *cos* (270°) = 0, while it completely opposes capillarity at *cos* (180°) = −1. The juxtaposition illustrates the spatial limitations of Nsf. Moreover, lateral rapid flow requires saturated conditions along a layer with path widths narrower than F.

Pressure in the WCW is atmospheric while $\psi < 0$ typically prevails ahead of it. Therefore, water is abstracted from the WCW onto L. Abstraction is usually completed during short periods. The amount of abstraction shows in the difference between θ_{end}, when the WCW has ceased, and θ_{ante} before the passing of the WCW. θ_{end} is approached with Equation (8) for a particular depth Z, while setting the WCW's end at $w(Z,t)/w_S = u$, where $u << 1$ is an arbitrarily selected threshold that depends on the particular application of Newton's shear flow. Water abstracted from the WCW is thus available for redistribution and uptake by plant roots.

4.4. Scales

The process scale during Nsf with constant F and L follows from the time and depth of interception, T_I and Z_I, Equations (6) and (7) that depend on the duration $(T_E - T_B)$ of the input and on F. The velocity of the wetting shock front reduces after T_I and beyond Z_I according to Equation (9) [48].

The system-related scales of the applicability of Nsf range from a couple of sand grains at the mm-scale to the km-scale. From neutron radiographs taken during infiltration in sand boxes, Hincapié and Germann (2010) [56] calculated fluxes in layers that were about 1 mm thick and volume balances in the cm^3-range. Flow across 21 vertical m of an ancient sand dune was already presented above [49], while tracer experiment across crystalline rocks suggests the applicability of Newton's shear flow approach to the km-range. Dubois (1991) [57] injected the tracers uranin and eosin about 1800 m above the Mont Blanc car tunnel that connects Chamonix (France) with Courmayeur (Italy). Within 108 d he detected the tracer fronts in seeps in the car tunnel. This amounts to $v_W \approx 2 \times 10^{-4}$ m s^{-1} that is well within the range of the v_W-collection reported above. Moreover, the gentle water seeps in the tunnel revealed that gravity was consumed by viscosity during the long-lasting flow.

The observations of [57] across 1800 m of crystalline rocks of the Mont Blanc massif and the water balance calculations of finger flow in the sand boxes of [56] at the scale of millimeters hint at the spatial tolerance of Newton's shear flow. This may advance the approach to an attractive tool, for instance, for the study of infiltration into groundwater systems.

4.5. Time-Variable Infiltration

So far input to the surface is considered a single pulse $P(q_S, T_B, T_E)$. Time variable input needs to be separated into a series of rectangular pulses each carrying its individual parameters. Under the assumption of the macropore flow restriction i.e., $dL/dq = 0$, a series of pulses can be routed as a sequence of kinematic waves according [30], whose mathematical approach models Newton's shear flow correctly if $a = 3$ in Equation (12). [58] provides details of applying characteristics to the multi-pulse infiltration, while [49] lists some results.

Upon observations of infiltrations with varying durations and the rates from 5, to 10, 20, and 40 mm h^{-1} into a column of an undisturbed soil [59], postulated the macropore flow restriction of $dL/dq_S = 0$, meaning that flow occurs along the same paths independently from the input rate. They tested the relationship of $v_W\,(q_S) = b^{1/3} \times q_S{}^{2/3}$, Equations (1) to Equation (3), that resulted in an acceptable coefficient of determination of $r^2 = 0.95$. However, that was an exception hardly achieved again. This calls for further investigation of the relationship, most likely under consideration of θ_{ante}.

4.6. Transport of Tracers and Particles

Preferential flow in soil hydrology is frequently associated with enhanced and accelerated particle, solute and, primarily, pollutant breakthrough [60]. However, Bogner and Germann (2019) [61] report considerable delays of tracer breakthrough compared with the first arrival of wetting shock fronts at the drain of soil columns with heights of 0.4 m. They referred to the phenomenon as 'pushing out old water' that is well known in catchment hydrology. They statistically explained 81% of the observed delay variations with combinations of L and F when applying Newton's shear flow to the data. It appears that tracer exchange on large L from thin F of the WCW may be even faster than presumed 'preferential' tracer transport. Under consideration of the mechanistic parameters F and L,

Newton's shear flow provides for a novel tool for the unambiguous investigation of tracer transport and exchange i.e., accelerated as well as decelerated breakthroughs that primarily addresses the tracer's mass balance and secondarily diffusion.

Similar considerations may apply to the transport of particles and microbes. Based on a precursor of Nsf, [62] approached with drag-forces the transport of latex beads and bacteriophages in soils. The three types of electrically uncharged latex beads had dimeters of 0.5, 1.0, and 1.75 μm. Maximal particle concentrations relative to the input suspensions reached about 0.003. While the relationships between relative particle concentrations vs. drag force showed intermediate to strong linearity during the increasing and decreasing limbs of the drain hydrograph, the drag force in the trailing was apparently too small to produce a considerable effect on their translocations. The five bacteriophages H40/l, H4/4, T7, H6/l, and fl carried ζ-potentials in the range of −50.1 to −31.7 mV. Their much weaker relative concentrations vs. drag force showed pronounced, yet undisclosed hysteresis. Recently, [63] called for improving our understanding of the transport of microplastic particles in soils, while [64] found these particles in worm burrows and assumed preferential flow along macropores as the major transport process. Thus, it might be worthwhile to further develop Newton's shear flow application to the particle transport under particular attention of the films thickness F and their specific contact areas L.

4.7. Ecohydrology

The early 20th century saw numerous runoff studies in headwater catchments, many of them with the purpose of demonstrating the advantage of forests over open lands in the mitigation of flood and debris flows from steep slopes, mainly to convince governments for subsidizing huge reforestation projects. Burger's (1922) [22] investigations with soil cores as well as his comparative runoff measurements from the 50% forested watershed of Rappengraben vs. the completely forested catchment of Sperbelgraben are well known in forest hydrology. However, not all the compartments from precipitation to runoff got the necessary attention in research they would require for closing the water balance. For instance, the infiltration experiments of [22] were not carried through to assess with them drainage and runoff in the headwater catchments.

Issues of reforestations' remedial effects on poorly permeable clay soils at sites in the Swiss pre-alps have surfaced again because numerous sites require rejuvenation of the more than 100 y old stands. In this context, Lange et al. (2009) [65] were able to demonstrate positive effects of tree root density on rapid infiltration in stagnic soils.

Wiekenkamp et al. (2019) [66] compared in the Wüstebach (BRD) catchment the hydrology of a forest site with the neighboring clear-cut site. They concluded that infiltration via preferential flow paths increased after deforestation. However, their purposeful exclusion of interception may deprive them of further interpreting the difference between the two treatments in rapid infiltration because interception may turn out as major gate controlling the input pulses $P(q_S,T_B,T_E)$.

5. Summary and Conclusions

The first section reports from English chalk rapid contaminant transport about 150 to 700 faster than the slow tracer movements. The second section deals with the evolution of infiltration-drainage concepts since the mid-19th Century. Two lines of thought emerge: On the one hand, Schumacher (1864) [10] suggested a two-processes approach for flow and storage of water in soils, one accounting for infiltration, the other one for capillary flow. On the other hand, a dual-porosity approach follows from the observations of Lawes et al. (1882) [11]. In searching an universal approach for storage and flow of water in partially saturated soils, Buckingham (1907) [12] and Richards (1931) [3] focused on capillarity and the soil hydrological functions K-θ-ψ. Concepts of non-equilibrium flow emerged upon discovered discrepancies between theory and observations that led later on to dual-porosity approaches. The third section takes off from Schumacher's two-processes approach, exploring fast infiltration and drainage as gravity-driven and viscosity controlled flow, resulting in a water content wave, *WCW*, that is based on Newton's shear flow i.e., laminar flow in unsaturated permeable media lasting two to ten times longer than the duration of input and occurring under atmospheric pressure. This restriction suggests strong capillary gradients between the *WCW* and the sessile parts. Because of the wide specific contact area, expressed with the factor L in the typical range of 500 to 20,000 $m^2\ m^{-3}$, and the thin films F that are typically between 2 and 50 μm thick, exchanges of water, energy, heat, solutes, and particles are fast and strong. The fourth section elucidates applications of the approach.

On the one side, the universal Richards equation deals with infiltration and redistribution of soil water. It assumes that entire soil moisture θ participates in the flow process, thus inevitably leading to apparent conditions of non-equilibrium, while dual-porosity approaches are hoped to lead away from the uneasy situation. On the other side, Newton's shear flow focuses on viscosity and gravity resulting in adequate time scales during infiltration and associated drainage. Once the wetting shock fronts have ceased to advance, capillarity takes over and soil moisture redistributes towards ψ-θ equilibria. This leads to a dual-process approach to infiltration and redistribution, where time and depth of front interception, T_I and Z_I, serve as scales for separating the two processes. A further consequence of concentrating flow and transport on Newton's shear flow are the spatio-temporal limits of the processes, expressed with a few multiples of T_I and Z_I. Moreover,' pushing out old water' and assuming flow rates under atmospheric pressure exceeding K_{sat} indicate novel aspects associated with Newton's infiltration that were not considerable in previous approaches to preferential flows. Finally, the analytical expressions are amenable to mathematical procedures, such as kinematic wave theory, and their theoretical combinations may lead to new and solid hypotheses calling for experimental testing.

Funding: This research did not receive any external funding

Acknowledgments: I thank Scholle, Guest Editor of the Special Issue on "Physical and Mathematical Fluid Mechanics", for inviting the manuscript to WATER. I deeply appreciate the positive critiques of three anonymous reviewers that helped considerably to straighten out the manuscript.

Conflicts of Interest: The author declares no conflict of interest.

References

1. Reeves, M.J. Recharge of the English chalk, A possible mechanism. *Eng. Geol.* **1980**, *14*, 231–240. [CrossRef]
2. Atkinson, T.C.; Smith, D.I. Rapid groundwater flow in fissures in the chalk: An example from South Hampshire. *Q. J. Eng. Geolog.* **1974**, *7*, 197–2016. [CrossRef]
3. Richards, L.A. Capillary conduction of liquids through porous mediums. *Physics* **1931**, *1*, 318–333. [CrossRef]
4. Jarvis, N.; Koestel, J.; Larsbo, M. Understanding Preferential Flow in the Vadose Zone: Recent Advances and Future Prospects. *Vadose Zone J.* **2016**, 15. [CrossRef]
5. Morbidelli, R.; Corradini, C.; Saltalippi, C.; Flammini, A.; Dari, J.; Govindaraju, R.S. Rainfall infiltration modelling. *Water* **2018**, *10*, 1873. [CrossRef]
6. Poiseuille, J.L.M. Recherches expérimentales sur le mouvement des liquides dans les tubes de très petits diamètres. *Comptes Rendus, xi-xii, Mém. des Sav. Etrangers* **1846**, *ix*, 2.
7. Darcy, H. *Les Fontaines Publiques de la Ville de Dijon*; Dalmont: Paris, France, 1856.
8. Dupuit, J. *Études Théoriques et Pratiques sur le Movement des Eaux Dans les Canaux Découverts et à Travers les Terrains Permeable*, 2nd ed.; Dunod: Paris, France, 1863.
9. Forchheimer, P.H. Wasserbewegung durch Boden. *Zeitschr. Ver. Deutscher Ingenieure* **1901**, *45*, 1782–1788.
10. Schumacher, W. *Die Physik des Bodens*; Wiegandt & Hempel: Berlin, Germany, 1864.
11. Lawes, J.B.; Gilbert, J.H.; Warington, R. *On the Amount and Composition of the Rain and Drainage Water Collected at Rothamsted*; Williams, Clowes and Sons Ltd.: London, UK, 1882.
12. Buckingham, E. *Studies on the Movement of Soil Moisture*; Bulletin 38; U.S. Department of Agriculture, Bureau of Soils: Washington, DC, USA, 1907.
13. Young, T. An essay on the cohesion of fluids. *Philos. Trans. R. Soc. Lond.* **1805**, *95*, 65–87.
14. Fourier, J.B.J. *Théorie Analytique de la Chaleur*; Firmin-Didot: Paris, France, 1822.

15. Ohm, G.S. Vorläufige Anzeige des Gesetzes, nach welchem Metalle die Contact Eletricität leiten. *Annalen der Physik und Chemie.* **1825**, *80*, 79–88. [CrossRef]
16. Or, D. The tyranny of small scales on presenting soil processes in global land surface models. *Water Resour. Res.* **2018**. [CrossRef]
17. Richardson, L.F. *Weather Prediction by Numerical Process*; Cambridge University Press: Cambridge, UK, 1922.
18. Gardner, W.; Israelsen, O.W.; Edlefsen, N.E.; Clyde, H. The capillary potential function and its relation to irrigation practice. *Phys. Rev.* **1922**, *20*, 196.
19. Philip, J.R. Theory of infiltration. *Adv. Hydrosci.* **1969**, *5*, 215–290.
20. van Genuchten, M.T. A Closed-form Equation for Predicting the Hydraulic Conductivity of Unsaturated Soils. *Soil Sci. Soc. Am. J.* **1980**, *44*, 892–898. [CrossRef]
21. Simunek, J.; Senja, M.; van Genuchten, M.T. *HYDRUS-2D, Simulating water flow, heat and solute transport in two-dimensional variably saturated media*; Version 2.0. USDA-ARS; International Ground Water Modeling Center: IGMC-TPS53; U.S. Salinity Laboratory: Riverside, CA, USA; Colorado School of Mines: Golden, CO, USA, 1999.
22. Burger, H. Physikalische Eigenschaften von Wald- und Freilandböden. *Mitt. Schweiz. Centralanstalt Forstl. Vers'wesen* **1922**, *XIII*, 1–221.
23. Veihmeyer, F.J. Some factors affecting the irrigation requirements of deciduous orchards. *Hilgardia* **1927**, *2*, 190. [CrossRef]
24. Veihmeyer, F.J. Soil Moisture. In *Handbuch der Pflanzenphysiologie*; Springer: Verlag, Heidelberg, 1954.
25. Bouma, J.; Jongerius, A.; Bursman, O.; Jager, A.; Schonderbeek, D. The function of different types of macropores during saturated flow through four swelling soil hirizons. *Soil Sci. Sco. Am. J.* **1977**, *41*, 945–950. [CrossRef]
26. Beven, K.; Germann, P. Macropores and water flow in soils. *Water Resour. Res.* **1982**, *18*, 1311–1325.
27. Abou Najm, M.; Lassabatere, L.; Stewart, R.D. Current insights into nonuniform flow across scales, processes, and applications. *Vadose Zone J.* **2018**, *18*.
28. Beven, K. A century of denial: Preferential and nonequilibrium water flow in soils, 1864–1984. *Vadose Zone J.* **2018**, *17*, 180153. [CrossRef]
29. Atalah, N.M.; Abou Najm, M. Characterization of synthetic porous media using non-Newtonian fluids: Experimental evidence. *Europ. J. Soil Sci.* **2018**. [CrossRef]
30. Newton, I. *The Mathematical Principles of Natural Philosophy—Translation into English*; Benjamin Motte: London, UK, 1729; Volume II, p. 184.
31. Germann, P.; Beven, K. Water flow in soil macropores. I. an experimental approach. *J. Soil Sci. (British)* **1981**, *82*, 1–13.
32. Germann, P.; Beven, K. Water flow in soil macropores. III. A statistical approach. *J. Soil Sci. (British)* **1981**, *82*, 31–39. [CrossRef]
33. Beven, K.; Germann, P. Water flow in soil macropores. II. A combined flow model. *J. Soil Sci.(British)* **1981**, *32*, 15–29.
34. Lighthill, M.J.; Whitham, G.B. On kinematic waves I. Flood movement in long rivers. *Proc. R. Soc. London. Ser. A. Math. Phys. Sci.* **1955**, *229*, 281–316.
35. Germann, P. Kinematic wave approach to infiltration and drainage into and from soil macropores. *Trans. ASAE* **1985**, *28*, 745–749. [CrossRef]
36. Germann, P.; Edwards, W.M.; Owens, L.B. Profiles of Bromide and Increased Soil Moisture after Infiltration into Soils with Macropores. *Soil Sci. Soc. Am. J.* **1984**, *48*, 237–244. [CrossRef]
37. Germann, P. Rapid drainage response to precipitation. *Hydrol. Proc.* **1986**, *1*, 3–13. [CrossRef]
38. Germann, P.; DiPietro, L.; Singh, V. Momentum of flow in soils assessed with TDR-moisture readings. *Geoderma* **1997**, *80*, 153–168. [CrossRef]
39. Flammer, I.; Blum, A.; Leiser, A.; Germann, P. Acoustic assessment of flow patterns in unsaturated soil. *J. Appl. Geophys.* **2001**, *46*, 115–128. [CrossRef]
40. Germann, P. Viscosity-the weak link between Darcy's law and Richards' capillary flow. *Hydrol. Proc.* **2018**, *32*, 1166–1172. [CrossRef]
41. Germann, P.; Beven, K. Kinematic wave approximation to infiltration into soils with sorbing macropores. *Water Resour. Res.* **1985**, *21*, 990–996. [CrossRef]

42. Germann, P.; Helbling, A.; Vadilonga, T. Rivulet Approach to Rates of Preferential Infiltration. *Vadose Zone J.* **2007**, *6*, 207–220. [CrossRef]
43. Kutilek, M.; Germann, P. Converging hydrostatic and hydrodynamic concepts of preferential flow definitions. *J. Contam. Hydrol.* **2009**, *104*, 61–65. [CrossRef] [PubMed]
44. Germann, P. *Preferential Flow—Stokes Approach to Infiltration and Drainage*; University of Bern: Bern, Switzerland, 2014; Available online: https://boris.unibe.ch/119081/1/preferential_flow.pdf (accessed on 22 January 2020).
45. Germann, P. Preferential flow at the Darcy scale: Parameters from water content time series. *Methods Soil Anal.* **2018**, *3*, 160121. [CrossRef]
46. Germann, P.; Karlen, M. Viscous-flow approach to in situ infiltration and in vitro saturated hydraulic conductivity determination. *Vadose Zone J.* **2016**, *15*. [CrossRef]
47. Selker, J.; Parlange, J.-Y.; Steenhuis, T. Fingered flow in two dimensions: 2. Predicting finger moisture profile. *Water Resour. Res.* **1992**, *28*, 2523–2528. [CrossRef]
48. Germann, P.; al Hagrey, S.A. Gravity-driven and viscosity-dominated infiltration in a full-scale sand model. *Vadose Zone J.* **2008**, *7*, 1160–1169. [CrossRef]
49. Germann, P.; Prasuhn, V. Viscous flow approach to rapid infiltration and drainage in a weighing lysimeter. *Vadose Zone J.* **2018**, *17*, 170020. [CrossRef]
50. Rimon, Y.; Dahan, O.; Nativ, R.; Geyer, S. Water percolation through a deep vadose zone and groundwater recharge: Results based on a new vadose zone monitoring system. *Water Resour. Res.* **2007**, *43*, W05402. [CrossRef]
51. Hincapié, I.A.; Germann, P. Abstraction from Infiltrating Water Content Waves during Weak Viscous Flows. *Vadose Zone J.* **2009**, *8*, 996–1003. [CrossRef]
52. Demand, D.; Blume, T.; Weiler, M. Spatio-temporal relevance and controls of preferential flow at the landscape scale. *Hydrol. Earth Syst. Sci.* **2019**, *23*, 4869–4889. [CrossRef]
53. Blum, A.; Flammer, I.; Friedly, T.; Germann, P. Acoustic tomography applied to water flow in unsaturated soils. *Vadose Zone J.* **2004**, *3*, 288–299. [CrossRef]
54. Germann, P.; Jäggi, E.; Niggli, T. Rate, kinetic energy and momentum of preferential flow estimated from in situ water content measurements. *Europ. J. Soil Sci.* **2002**, *53*, 607–617. [CrossRef]
55. Germann, P.; Hensel, D. Poiseuille flow geometry inferred from velocities of wetting fronts in soils. *Vadose Zone J.* **2006**, *5*, 867–876. [CrossRef]
56. Hincapié, I.; Germann, P. Water Content Wave Approach Applied to Neutron Radiographs of Finger Flow. *Vadose Zone J.* **2010**, *9*, 278–284. [CrossRef]
57. Dubois, J.-D. *Typologie des aquifers du cristallin: Exemples des massifs des Aiguilles Rouges et du Mont-Blanc.* (Typology of Aquifers in the Cristaline: Examples from the Massifs Aguilles Rouges and Mt. Blanc). Ph.D. Dissertation, Department of Civil Engineering, EPFL, Lausanne, Switzerland, 1991. Available online: https://www.swisstransfer.com/d/1442dd60-cf83-4220-a8e0-01e30bd5e34f (accessed on 22 January 2020).
58. Germann, P. Hydromechanics and kinematics in preferential flow. *Soil Sci.* **2018**, *183*, 1–10. [CrossRef]
59. Hincapié, I.; Germann, P. Impact of initial and boundary conditions on preferential flow. *J. Contam. Hydrol.* **2009**, *104*, 67–73. [CrossRef]
60. Larsbo, M.; Koestel, J.; Jarvis, N. Relations between macropore network characteristics and the degree of preferential solute transport. *Hydrol. Earth Syst. Sci.* **2014**, *18*, 5255–5269. [CrossRef]
61. Bogner, C.; Germann, P. Viscous flow approach to "pushing out old water" from undisturbed and repacked soil columns. *Vadose Zone J.* **2019**, *18*, 180168. [CrossRef]
62. Germann, P.; Alaoui, A.; Riesen, D. Drag force approach to the transport of colloids in unsaturated soils. *Water Resour. Res.* **2002**, *38*, 18-1. [CrossRef]
63. Bigalke, M.; Montserrat, F. Foreword to the research front on 'Microplasitcs in Soils'. *Environ. Chem.* **2019**, *16*, 1–2. [CrossRef]
64. Yu, M.; van der Ploeg, M.; Lwanga, E.H.; Yang, X.; Zhang, S.; Ma, X.; Ritsema, C.J.; Geissen, V. Leaching of microplastics by preferential flow in earthworm (*Lumbricus terrestris*) burrows. *Environ. Chem.* **2019**, *16*, 31–40. [CrossRef]

65. Lange, B.; Lüscher, P.; Germann, P.F. Significance of tree roots for preferential infiltration in stagnic soils. *Hydrol. Earth Syst. Sci.* **2009**, *13*, 1809–1821. [CrossRef]
66. Wiekenkamp, I.; Huisman, J.A.; Bogena, H.R.; Vereecken, H. Effects of deforestation on water flow in the vadose zone. *Water* **2020**, *12*, 35. [CrossRef]

Article

Poroacoustic Traveling Waves under the Rubin–Rosenau–Gottlieb Theory of Generalized Continua

Pedro M. Jordan

Acoustics Division, U.S. Naval Research Laboratory, Stennis Space Ctr., Hancock County, MS 39529, USA; pedro.jordan@nrlssc.navy.mil

Received: 16 November 2019; Accepted: 4 March 2020; Published: 14 March 2020

Abstract: We investigate linear and nonlinear poroacoustic waveforms under the Rubin–Rosenau–Gottlieb (RRG) theory of generalized continua. Working in the context of the Cauchy problem, on both the real line and the case with periodic boundary conditions, exact and asymptotic expressions are obtained. Numerical simulations are also presented, von Neumann–Richtmyer "artificial" viscosity is used to derive an exact kink-type solution to the poroacoustic piston problem, and possible experimental tests of our findings are noted. The presentation concludes with a discussion of possible follow-on investigations.

Keywords: poroacoustics; Rubin–Rosenau–Gottlieb theory; solitary waves and kinks

---

1. Introduction

What is known today as the "RRG theory" was put forth by Rubin et al. [1] in 1995. This phenomenological-based theory of generalized continua is thought capable of modeling dispersive effects caused by the introduction of a medium's characteristic length, which Rubin et al. denote as α. Under RRG theory, α is regarded as an inherent material property. From the modeling standpoint, this theory exhibits a number of appealing features, the two most important of which are the following: (i) it is only the pressure stress (i.e., isotropic) part of the Cauchy (i.e., total) stress tensor and the specific Helmholtz free energy that are modified, but these modifications are achieved by adding perburtative terms, which must satisfy certain constraint equations, to the constitutive relations of the former and latter; and (ii), no additional boundary nor initial conditions, beyond those required to solve classically formulated problems, are needed ([1], p. 4063).

To date, RRG theory has only been applied to single-phase media; see, e.g., Ref. [2] and those cited therein. Hence, there is an obvious need to investigate the nature of the solutions, e.g., those of the traveling wave type, predicted by this theory in the case of multi-phase media.

Accordingly, the aim of this communication is to carry out a *preliminary* investigation of RRG theory in the context of acoustic problems involving propagation in dual-phase (specifically, fluid + solid) media—dual-phase media being, of course, the simplest case of multi-phase media. Employing both analytical and numerical methodologies, we consider linear and finite-amplitude poroacoustic propagation under the RRG-based generalization of what some refer to as the *Brinkman poroacoustic model* (BPM) (Although he does not refer to it as such, the general, multi-D, version of the BPM follows on setting $C = 0$ in Burmeister [3].). Here, it should be noted that the original version of the drag law on which the BPM is based reads (see, e.g., Refs. [4,5])

$$\nabla P = \tilde{\mu}\chi\nabla^2\mathbf{u} - (\mu\chi/K)\mathbf{u}. \tag{1}$$

In this relation, $\mathbf{u}$ is the intrinsic average velocity of the fluid, which it is related to $\mathbf{v}$, the Darcy velocity, via the Dupuit–Forchheimer relationship $\mathbf{v} = \chi\mathbf{u}$ ([4], p. 5); P is a pressure, an intrinsic quantity, which is not the thermodynamic pressure ([4], §1.4.1); μ is the usual shear viscosity coefficient; $\tilde{\mu}$ is an effective viscosity ([4], §1.5.3), which is often referred to as the *Brinkman viscosity*; and $K > 0$ and $\chi \in (0,1)$, the permeability and porosity of the solid matrix, are assumed to be constants. We should also note that Equation (1) reduces to Darcy's law on setting $\tilde{\mu} := 0$.

Before beginning our analysis, we should point out that when traveling wave solutions (TWS)s in the form of kinks are encountered below, their shock thicknesses shall be expressed using Prandtl's definition (see, e.g., (Ref. [6], p. 680)), viz.:

$$\text{shock thickness} := \frac{\mathfrak{F}(-\infty) - \mathfrak{F}(+\infty)}{\max_{\mathfrak{z}\in\mathbb{R}} |\mathrm{d}\mathfrak{F}(\mathfrak{z})/\mathrm{d}\mathfrak{z}|}, \tag{2}$$

where $\mathfrak{z}$ represents the wave (i.e., similarity) variable. Herein, all traveling wave profiles shall be taken to be propagating to the right along the axis corresponding to the wave variable under consideration.

2. Mathematical Formulations

2.1. Poroacoustic Model Systems

When entropy production in the fluid, which we hereafter take to be a *perfect gas* [7], is ignored (i.e., we take the flow to be *homentropic* ([7], p. 60)) and the porous matrix is regarded as being both stationary and composed of a thermally non-conducting rigid solid, the 1D versions of the RRG-based model we propose *and* the BPM become, in the case of propagation along the x-axis,

$$\varrho_t + u\varrho_x + \varrho u_x = 0, \tag{3a}$$

$$\varrho(u_t + uu_x) = \tilde{\mu}\chi u_{xx} - (\mu\chi/K)u - \begin{cases} \wp_x, & \text{BPM}, \\ [\wp - 2\alpha^2\varrho(u_{xt} + uu_{xx})]_x, & \text{RRG}, \end{cases} \tag{3b}$$

$$\wp \approx \wp_0(\varrho/\varrho_0)^\gamma \qquad (\mathfrak{n} \approx \mathfrak{n}_0). \tag{3c}$$

In System (3), $\wp(> 0)$ is the thermodynamic pressure; $\varrho(> 0)$ is the mass density of the gas; $\mathfrak{n}$ is the specific entropy of the gas; the parameter γ denotes the ratio of specific heats, where $\gamma \in (1, 5/3]$ in the case of perfect gases; we take $\alpha(> 0)$, which carries the unit of length, to be a constant (That is, we have assumed the *simplest* version of RRG theory; see (Ref. [1], Equation (20)).); the problem geometry dictates that, here and henceforth, $\mathbf{u} = (u(x,t), 0, 0)$, $\wp = \wp(x,t)$, and $\varrho = \varrho(x,t)$; and a zero ("0") subscript attached to a dependent variable denotes the (constant) equilibrium state value of that variable, where we note that $\mathbf{u}_0 = (0,0,0)$.

Here, we observe that since the flow has been assumed homentropic, our RRG-based poroacoustic model is obtained by perturbing *only* the pressure tensor term in the BPM. Also, we record for later reference that $c_0 = \sqrt{\gamma\wp_0/\varrho_0}$ is the (constant) equilibrium state value of the sound speed, i.e., the speed of sound in the undisturbed gas; see, e.g., (Ref. [7], §4.3).

2.2. Finite-Amplitude Equation of Motion: The Case $\mu :=$ const.

We begin this sub-section with the following observation: Because $\nabla \times \mathbf{u} = (0,0,0)$ holds identically under the present (1D) geometry, it follows that $\mathbf{u} = \nabla\phi$; therefore, $u(x,t) = \phi_x(x,t)$, where ϕ_x denotes the scalar velocity potential.

Hence, on invoking the finite-amplitude approximation, and introducing the following dimensionless variables:

$$u^\diamond = u/U_\mathrm{p}, \quad s = (\varrho - \varrho_0)/\varrho_0, \quad \phi^\diamond = \phi/(U_\mathrm{p}L), \quad x^\diamond = x/L, \quad t^\diamond = t(c_0/L), \tag{4}$$

where the positive constants L and U_p respectively denote a macro-length scale characteristic of the propagation domain and the magnitude of the peak particle velocity in the gas, it is not difficult to establish (See, e.g., the derivation performed in (Ref. [8], §2), and note that (Ref. [8], Equation (10)) is the $\sigma, \delta := 0$ special case of Equation (5) herein.) that the $\mu :=$ const. case of System (3) reduces to the weakly-nonlinear, *bi-directional*, equation of motion (EoM)

$$\phi_{tt} - [1 - 2\epsilon(\beta - 1)\phi_t]\phi_{xx} + \epsilon\partial_t(\phi_x)^2 = \sigma\phi_{txx} + a_0^2\phi_{ttxx} - \delta\phi_t, \tag{5}$$

where here and henceforth all diamond ($^\diamond$) superscripts have been suppressed for convenience. In Equation (5), which we note reduces to the corresponding EoM for the (1D) BPM on setting $a_0 := 0$, $\epsilon = U_p/c_0$ is the Mach number, where $\epsilon \ll 1$ is assumed; $\delta = \nu\chi L/(c_0 K)$ is the dimensionless Darcy coefficient, where $\nu = \mu/\varrho_0$ is the kinematic viscosity of the gas; a_0, the dimensionless version of α, is given by $a_0 = \alpha\sqrt{2}/L$; we have set $\sigma := \chi/Re_B$, where $Re_B = c_0 L/\tilde{\nu}$ is a Reynolds number, and where $\tilde{\nu} = \tilde{\mu}/\varrho_0$; and $\beta(> 1)$ denotes the *coefficient of nonlinearity* [9], which in the case of a perfect gas is given by

$$\beta = (\gamma + 1)/2. \tag{6}$$

In deriving Equation (5) we have assumed that $\delta, \sigma, a_0, |s| \sim \mathcal{O}(\epsilon)$ and, in accordance with the finite-amplitude approximation, only *nonlinear* terms $\mathcal{O}(\epsilon^2)$ have been neglected.

2.3. *Right-Running Equations of Motion for the Case $\mu :=$ const.*

Although derived under the finite-amplitude approximation, Equation (5) is still too complicated for treatment by analytical means. Fortunately, however, the nature of the problems to be considered below is such that we may employ the uni-directional approximation to reduce the order of Equation (5) by one and confine its nonlinearity to a single (quadratic) term. Omitting the details, we find that under, say, the right-running case of this approximation (See, e.g., Crighton's ([10], p. 16) derivation of the acoustic version of Burgers' equation.), which in the present setting reads $\phi_x \simeq -\phi_t$, our EoM becomes, after making use of the relation $u(x,t) = \phi_x(x,t)$ and simplifying,

$$u_t + u_x + \epsilon\beta u u_x - \tfrac{1}{2}a_0^2 u_{txx} + \tfrac{1}{2}\delta u = \tfrac{1}{2}\sigma u_{xx}, \tag{7}$$

which on switching to the variables $X = x - t$ and $T = t$ is further reduced to

$$u_T + \epsilon\beta u u_X - \tfrac{1}{2}a_0^2 u_{TXX} + \tfrac{1}{2}\delta u = \tfrac{1}{2}\sigma u_{XX}. \tag{8}$$

If we once again make use of the right-running approximation, which now takes the form $u_T \simeq -u_X$, to re-express *only* the third order term in Equation (8), which is justified since $a_0 \sim \mathcal{O}(\epsilon)$ (i.e., $(a_0^2/2)u_{TXX}$ is a "small" term), then Equation (8) assumes its final form, specifically,

$$u_T + \epsilon\beta u u_X + \tfrac{1}{2}a_0^2 u_{XXX} + \tfrac{1}{2}\delta u = \tfrac{1}{2}\sigma u_{XX}, \tag{9}$$

a PDE which we term the *damped Burgers–KdV* (dBKdV) equation.

In closing this sub-section we stress that Equations (7)–(9) apply only to right-running waveforms; i.e., to problems wherein reflection (to the left) is not possible.

3. Comparison of Linearized EoMs: The Cauchy Problem

In this section we compare the BPM with its RRG-based counterpart under the linear approximation, which at the EoM level corresponds to setting $\epsilon := 0$. We do so in the context of what is perhaps the best known problem from classical PDE theory.

To this this end, we consider the linearized version of Equation (9) in the setting of the following initial value problem (IVP), i.e., in the setting of the classical *Cauchy problem*:

$$u_T + \tfrac{1}{2}a_0^2 u_{XXX} + \tfrac{1}{2}\delta u = \tfrac{1}{2}\sigma u_{XX}, \quad X \in \mathbb{R}, T > 0, \tag{10a}$$

$$u(X,0) = f(X), \quad X \in \mathbb{R}. \tag{10b}$$

Here, we take $f(X)$, our initial condition (IC), to be defined on the real line and such that its Fourier transform exists.

On applying the Fourier transform to both Equation (10a) and the IC, and then solving the resulting (first order) ODE, it is readily shown that

$$\hat{u}(k,T) = \hat{f}(k)\exp\left[-\tfrac{1}{2}\left(\delta + \sigma k^2 - \mathrm{i}a_0^2k^3\right)T\right], \tag{11}$$

where k is the Fourier transform parameter and a hat over a quantity denotes the Fourier transform of that quantity. In turn, applying $\mathcal{F}^{-1}(\cdot)$, the inverse Fourier transform, to Equation (11) gives

$$u(X,T) = \frac{1}{2\pi}\exp(-\delta T/2)\int_{-\infty}^{\infty}\hat{f}(k)\exp\left[-\tfrac{1}{2}\left(\sigma k^2 - \mathrm{i}a_0^2k^3\right)T\right]\exp(\mathrm{i}kX)\,\mathrm{d}k \tag{12}$$

3.1. The RRG Case: $a_0 > 0$

Using the convolution theorem, and letting $\mathrm{Ai}(\cdot)$ denote the Airy function of the first kind, the RRG (i.e., $a_0 > 0$) case of Equation (12) can be recast in the more explicit form

$$\begin{aligned} u(X,T) = \exp(-\delta T/2)\left(\tfrac{2}{3a_0^2T}\right)^{1/3} \\ \times\int_{-\infty}^{\infty}\mathcal{F}^{-1}\left[\hat{f}(k)\exp\left(-\tfrac{1}{2}\sigma k^2\right)\right]\mathrm{Ai}\left[(X-\mathcal{Y})\left(\tfrac{2}{3a_0^2T}\right)^{1/3}\right]\mathrm{d}\mathcal{Y} \quad (T>0). \end{aligned} \tag{13}$$

For obvious reasons, the following two special cases of $f(X)$ are of particular interest:

$$\begin{aligned} u(X,T) = \exp(-\delta T/2)\left(\frac{2}{3a_0^2T}\right)^{1/3} \\ \times\begin{cases}\frac{1}{\sqrt{2\pi\sigma T}}\int_{-\infty}^{\infty}\exp\left(-\tfrac{1}{2\sigma}\mathcal{Y}^2/T\right)\mathrm{Ai}\left[(X-\mathcal{Y})\left(\tfrac{2}{3a_0^2T}\right)^{1/3}\right]\mathrm{d}\mathcal{Y}, & f(X)=\mathfrak{d}(X) \\ \frac{1}{\sqrt{1+b\sigma T}}\int_{-\infty}^{\infty}\exp\left[-\tfrac{1}{2}b(1+b\sigma T)^{-1}\mathcal{Y}^2\right]\mathrm{Ai}\left[(X-\mathcal{Y})\left(\tfrac{2}{3a_0^2T}\right)^{1/3}\right]\mathrm{d}\mathcal{Y}, & f(X)=\mathrm{e}^{-bX^2/2}\end{cases} \quad (T>0). \end{aligned} \tag{14}$$

Here, $\mathfrak{d}(\cdot)$ denotes the Dirac delta function and $b(>0)$ is a (dimensionless) constant.

3.2. The BPM Case: $a_0 := 0$

If we assume instead the BPM, then the solution of IVP (10) is readily obtained on setting $a_0 := 0$ in Equation (12); for the two aforementioned cases of $f(X)$, we find that

$$u(X,T) = \begin{cases}\left[\frac{\exp(-\delta T/2)}{\sqrt{2\pi\sigma T}}\right]\exp\left(-\tfrac{1}{2\sigma}X^2/T\right), & f(X)=\mathfrak{d}(X) \\ \left[\frac{\exp(-\delta T/2)}{\sqrt{1+b\sigma T}}\right]\exp\left[-\tfrac{1}{2}b(1+b\sigma T)^{-1}X^2\right], & f(X)=\mathrm{e}^{-bX^2/2}\end{cases} \quad (T>0). \tag{15}$$

3.3. Remarks: RRG vs. BPM

With regard to the Gaussian IC, the primary difference between the linearized RRG and BPM cases is that the pulse profile corresponding to the former instantly becomes oscillatory about the X-axis, due to the Airy function in its integrand, while that of the latter maintains, for all $T > 0$, the shape and strict positivity of the initiating Gaussian. The clearly contrasting behaviors exhibited by these two models should, therefore, allow researchers to experimentally determine which of the two best describes propagation in a given poroacoustic system.

4. Comparison of Right-Running, Weakly-Nonlinear, EoMs: Special Cases with $\mu :=$ const.

Before examining it in its most general form, and for the benefit of those readers who are not well acquainted with the intricacies of nonlinear evolution equations, it is instructive to first review selected special cases of Equation (9). The right-running models discussed in the next three sub-sections, all of which, it should be noted, have applications beyond poroacoustics, will each have a role to play in the analysis performed in Section 4.4.

4.1. *Case (I): Damped KdV (dKdV) Equation*

This case follows on setting $\sigma := 0$ (i.e., setting $\tilde{\mu} := 0$), under which Equation (9) reduces to

$$u_T + \epsilon\beta u u_X + \tfrac{1}{2}a_0^2 u_{XXX} + \tfrac{1}{2}\delta u = 0. \tag{16}$$

This PDE, we observe, is the RRG-modified version of the right-running Darcy–Jordan model; see Section 4.3 below.

Since we have assumed $\delta \ll 1$, applying the *Kryloff–Bogoliubov* asymptotic expansion method to the dKdV equation yields, as Ott and Sudan [11] have shown, the large-T expression

$$u(X,T) \sim \exp(-2\delta T/3)\,\mathrm{sech}^2\left[(\zeta/a_0)\sqrt{\tfrac{\epsilon\beta}{6}}\exp(-2\delta T/3)\right], \qquad T \sim \mathcal{O}(1/\delta), \tag{17}$$

where we have taken $N(0) = 1$ (see Ref. [11]), and we let

$$\zeta = X - \tfrac{\epsilon\beta}{2\delta}[1 - \exp(-\tfrac{2}{3}\delta T)]; \tag{18}$$

see also Ref. [12], as well as Ref. [13] (Ref. [13] contains a number of recently identified typographical errors; see Appendix A below.) and those cited therein. Equation (17) represents a damped, and *decelerating*, solitary waveform, and as such it cannot be a soliton in the classical sense [14]. Note, however, that the acoustic version of the classic soliton solution of the KdV equation (see Ref. [8]) is recovered as the limiting case

$$u(X,T) = \mathrm{sech}^2\left[a_0^{-1}(X - \epsilon\beta T/3)\sqrt{\tfrac{\epsilon\beta}{6}}\right] \qquad (\delta \to 0). \tag{19}$$

4.2. *Case (II): Damped Burgers' Equation*

This case, which corresponds to setting $a_0 := 0$, reads

$$u_T + \epsilon\beta u u_X + \tfrac{1}{2}\delta u = \tfrac{1}{2}\sigma u_{XX}. \tag{20}$$

Equation (20) is the right-running EoM stemming from the BPM, and in this context it has recently been investigated by Rossmanith and Puri [15].

As shown by Nimmo and Crighton [16], this generalization of Burgers' equation does not admit a linearizing (i.e., Cole–Hopf type) transform. As shown by Malfliet [17], however, its TWS, which assumes the form of a damped kink, is readily approximated. To the order expressed explicitly in Ref. [17], the TWS of Equation (20) is given by

$$\begin{aligned} u(X,T) \approx \tfrac{1}{2}\exp(-\delta T/2)[1 - Y(X,T)]\{1 + a_3(T)[1 + Y(X,T)]Y^3(X,T) \\ + a_5(T) \quad [1 + Y(X,T)]Y^5(X,T)\}. \end{aligned} \tag{21}$$

Here,

$$Y(X,T) := \tanh\left[\frac{2}{\lambda_B}\left(X - \frac{\epsilon\beta[1 - \exp(-\delta T/2)]}{\delta}\right)\right], \tag{22}$$

where $\lambda_B = 4\sigma/(\epsilon\beta)$ is the shock thickness exhibited by the TWS given below in Equation (25);

$$a_3(T) = -(1 - \exp(-\delta T/2))/3; \tag{23}$$

and

$$a_5(T) = -[128 - 160\exp(-\delta T/2) + \delta\lambda_B^2 \exp(-\delta T/2)/\sigma + 32\exp(-\delta T)]/240. \tag{24}$$

In Ref. [17], the parameter c, which herein has the value $c = 2/\lambda_B$, is defined so that Equation (21) yields the limiting case

$$u(X,T) = \tfrac{1}{2}\left\{1 - \tanh\left[\tfrac{2}{\lambda_B}\left(X - \tfrac{1}{2}\epsilon\beta T\right)\right]\right\} \qquad (\delta \to 0). \tag{25}$$

Equation (25) and λ_B are the TWS, which we note takes the form of a *Taylor shock*, and corresponding shock thickness, which was determined using Equation (2), respectively, admitted by the classic Burgers equation.

4.3. Case (III): Damped Riemann Equation

In the poroacoustic context, this case corresponds to the right-running version of the weakly-nonlinear *Darcy–Jordan model* (Also known as the *Jordan–Darcy model*; see Ciarletta and Straughan [18], as well as Straughan [5].) (DJM) [19]; specifically, the first order PDE [20]

$$u_T + \epsilon\beta u u_X + \tfrac{1}{2}\delta u = 0, \tag{26}$$

which follows on setting $a_0 := 0$ *and* $\sigma := 0$ in Equation (9).

In the setting of the Cauchy problem, the exact solution of the damped Riemann equation is readily determined; see, e.g., Crighton ([21], p. 196). In the particular case of Equation (26), this solution can be expressed as [20]

$$u(X,T) = u_0(\xi)\exp(-\delta T/2), \tag{27a}$$

$$\alpha^\star(X - \xi) = u_0(\xi)[1 - \exp(-\delta T/2)]. \tag{27b}$$

Here, $\xi = \xi(X,T)$ is the wave variable; $u_0(\xi)$ is the IC; and $\alpha^\star$, the critical amplitude value for acceleration waves under the DJM, is given by [19]

$$\alpha^\star = \frac{\delta}{2\epsilon\beta}. \tag{28}$$

For the particular case $u_0(X) = \cos(2\pi X)$, it can be shown (see, e.g., (Ref. [20], p. 3)) that under System (27)

$$T_B^* = -2\delta^{-1}\ln\left(1 - \frac{\alpha^*}{2\pi}\right). \tag{29}$$

If $T_B^* \in \mathbb{R}^+$, then T_B^* is the time at which the solution of the Cauchy problem involving Equation (26) suffers (finite-time) *gradient catastrophe* ([22], p. 36), where it is expected that $\alpha^* < 2\pi$ ($\Longrightarrow T_B^* \in \mathbb{R}^+$) in all cases of practical interest.

4.4. Numerical Results

Inspired by, and closely following, Zabusky and Kruskal's [14] analysis of the classic KdV equation, in this subsection we perform numerical experiments on Equation (9), and its special cases listed above as Cases (I) and (II), in the setting of the following initial-boundary value problem (IBVP) with *periodic* boundary conditions:

$$u_T + \epsilon\beta u u_X + \tfrac{1}{2}a_0^2 u_{XXX} + \tfrac{1}{2}\delta u = \tfrac{1}{2}\sigma u_{XX}, \quad |X| < 1, T > 0, \tag{30a}$$

$$u(-1,T) = u(1,T), \quad T > 0, \tag{30b}$$

$$u(X,0) = \cos(2\pi X), \quad |X| < 1. \tag{30c}$$

In (Ref. [14], Figure 1), snapshots of the evolution of the KdV's solution profile were displayed in units of (dimensionless) time T_B, where Zabusky and Kruskal used T_B to denote the "breakdown time" (i.e., the time at which finite-time gradient catastrophe occurs) of the solution to the Cauchy problem involving the classic (i.e., undamped) Riemann equation. In our analysis of IBVP (30), T_B^* shall play the role of T_B.

The graphs presented in Figures 1–3 were computed and plotted using MATHEMATICA (ver. 11.2). Except for the value of $\beta(= 1.2)$, which corresponds to diatomic gases (e.g., air) [9], all other parameter values were selected based on our desire to produce clear, illustrative, graphs and the need to satisfy the assumptions under which Equation (9) was derived.

From Figure 1 it is easy to see that, except for attenuation of the profile (caused by the Darcy term) and a slight phase shift, the dKdV profiles are qualitatively similar to those of the classic KdV equation in the setting of IBVP (30). And, as is also true in the case of the latter, reducing (resp. increasing) a_0 increases (resp. decreases) the number of pulses seen in Figure 1b.

In contrast, the plots shown in Figures 2 and 3 highlight the fact that, like that of the damped Burgers equation, the dBKdV profile suffers attenuation, again due to the Darcy term, and it also develops a "dull sawtooth" appearance as it shocks-up (to the right), but never breaks since $\sigma > 0$. More interesting, however, is the fact that for large-T, both the damped Burgers equation and dBKdV profiles are seen to *re-assume* the periodic form of the IC. As Figures 2c and 3c illustrate, both profiles evolve to become damped, and in the dBKdV case slightly phase-shifted (to the left), versions of the IC given in Equation (30c). This suggests that for sufficiently large values of T, one may employ the approximations $u(X,T) \approx u_{1,2}(X,T)$, where

$$u_{1,2}(X,T) := \begin{cases} \exp[-\mathrm{e}_1(T)]\cos[2\pi X + \psi_1(T)], & \text{dBKdV equation} \\ \exp[-\mathrm{e}_2(T)]\cos[2\pi X + \psi_2(T)], & \text{damped Burgers' equation} \end{cases} \qquad (T \gg T_B^*), \tag{31}$$

and where we require $\mathrm{e}_{1,2}(T) > 0$. Comparing the blue-broken curve in Figure 2c with its counterpart in Figure 3c we see that $\mathrm{e}_2(20T_B^*) > \mathrm{e}_1(20T_B^*) > 0$ while $\psi_1(20T_B^*) > \psi_2(20T_B^*) := 0$; here, for simplicity, we have *assumed* $\mathrm{e}_{1,2}(T)$ and $\psi_{1,2}(T)$ to be linear functions of T. In the setting of IBVP (30), then, the presence of the third order (i.e., RRG) term in the dBKdV equation gives rise to both a phase shift and slightly less attenuation vis-à-vis the damped Burgers equation.

While their usefulness may be limited to certain "windows" of T-values, the functions $\mathrm{e}_{1,2}(T)$ and $\psi_{1,2}(T)$ should be constructible based on Equation (31) and numerically generated, large-T, data sets using one of the many data-fitting methodologies found in the literature.

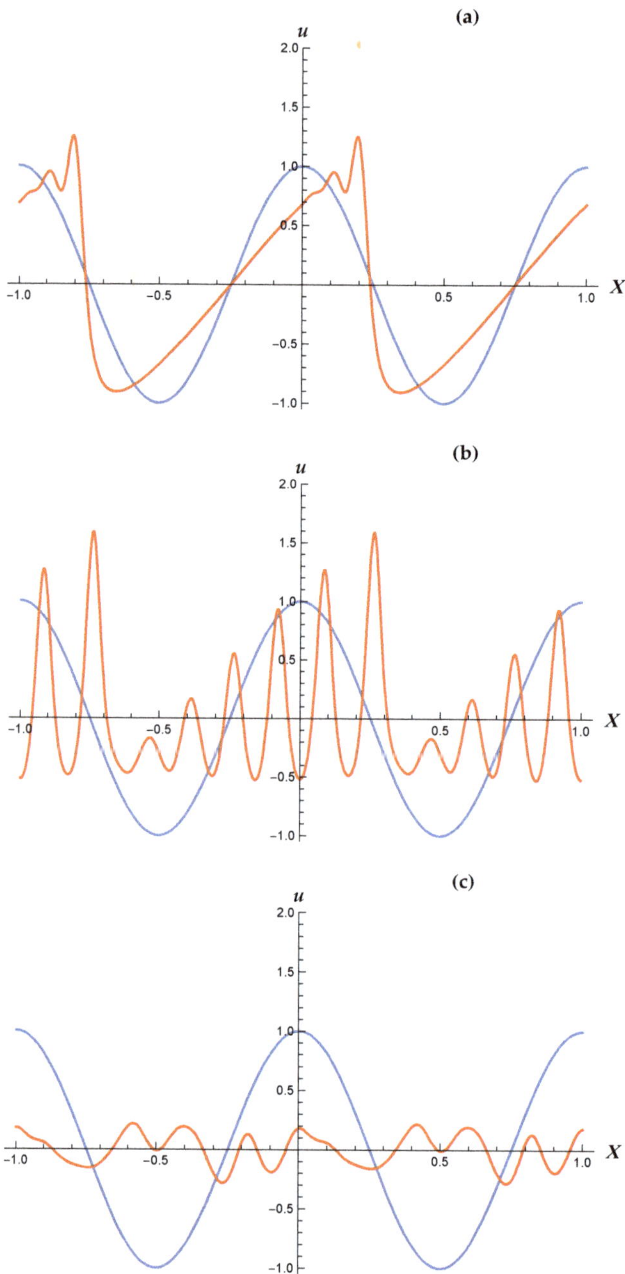

Figure 1. The dKdV case of IBVP (30). (**a**–**c**) correspond to $T = T_B^*$, $3.6T_B^*$, and $20T_B^*$, respectively, where $T_B^* \approx 1.382$. Red curves: u vs. X for $a_0 = (0.005)\sqrt{2}$, $\sigma = 0$, $\epsilon\beta = 0.12$, $\delta = 0.12$, and $\alpha^{\star} = 0.5$. Blue curves: IC given in Equation (30c).

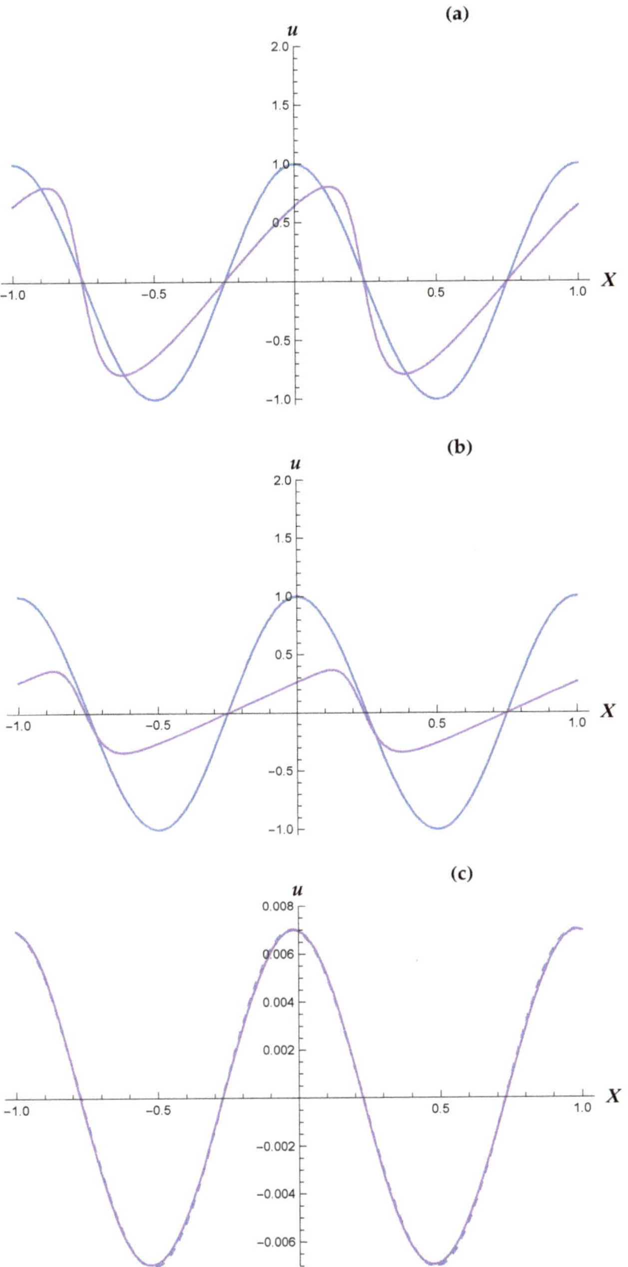

Figure 2. The dBKdV case of IBVP (30). (**a**–**c**) correspond to $T = T_B^*$, $3.6T_B^*$, and $20T_B^*$, respectively, where $T_B^* \approx 1.382$. Purple curves: u vs. X for $a_0 = (0.005)\sqrt{2}$, $\sigma = 0.005$, $\epsilon\beta = 0.12$, $\delta = 0.12$, and $\alpha^\star = 0.5$. Blue curves (solid): IC given in Equation (30c). Blue-broken curve: $u_1(X, 20T_B^*)$ vs. X (see Equation (31)), where we have set $\wp_1(20T_B^*) := (29.9)\delta T_B^*$ and $\psi_1(20T_B^*) := (0.1013)T_B^*$ based on a series of trial-and-error "visual fits".

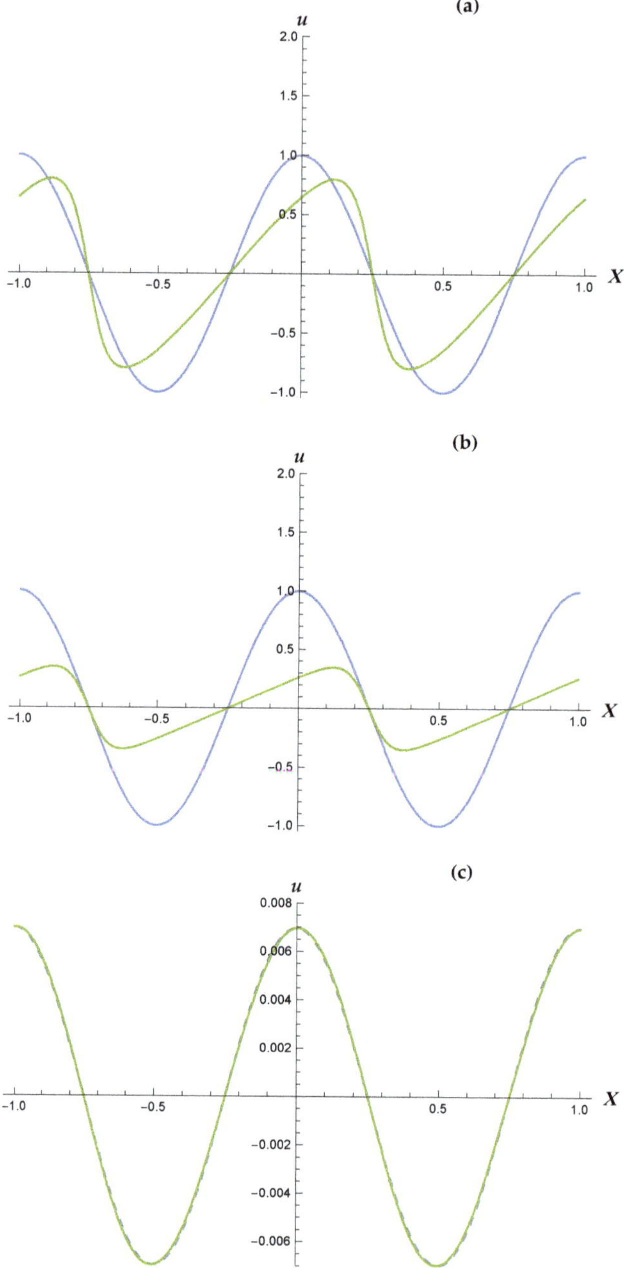

Figure 3. The damped Burgers equation case of IBVP (30). (**a–c**) correspond to $T = T_B^*$, $3.6T_B^*$, and $20T_B^*$, respectively, where $T_B^* \approx 1.382$. Green curves: u vs. X for $a_0 := 0$, $\sigma = 0.005$, $\epsilon\beta = 0.12$, $\delta = 0.12$, and $\alpha^{\star} = 0.5$. Blue curves (solid): IC given in Equation (30c). Blue-broken curve: $u_2(X, 20T_B^*)$ vs. X (see Equation (31)), where we have set $\mathrm{E}_2(20T_B^*) := (29.946)\delta T_B^*$ and $\psi_2(20T_B^*) := 0$ based on a series of trial-and-error "visual fits".

5. The RRG Case with "Artificial" μ

In 1950, von Neumann and Richtmyer (vNR) [23] introduced their artificial viscosity coefficient. In this section, we make use of this celebrated artifice not to regularize numerical schemes used to calculate shock profiles, as was vNR's aim, but rather to obtain an *analytical* solution to the poroacoustic version of the piston problem (Unlike Ref. [23], wherein Lagrangian coordinates were used, in this communication we employ the Eulerian description; see, e.g., (Ref. [24], §*V-D-1*) wherein vNR's system is recast under the latter.).

To this end, we return to the RRG case of System (3) and assume that

$$\mu \propto \alpha^2 \varrho |u_x|, \tag{32}$$

but continue to regard $\tilde{\mu}$ as a positive constant. Here, we have expressed the length-scale factor in vNR's artificial viscosity coefficient as α, instead of some grid spacing Δx.

For simplicity, we now assume that the porous solid in question is comprised of packed beds of rigid solid spheres, all of radius $r(>0)$, which are fixed in place. For such a configuration, the permeability is given by the well known *Kozeny–Carmen* relation [4]:

$$K = \frac{r^2 \chi^3}{45(1-\chi)^2}. \tag{33}$$

As these spheres are scatters of acoustic waves, we take α to be proportional to the characteristic length now associated with our dual-phase medium; i.e., we take $\alpha = b_1 r$, where $b_1(>0)$ is an "adjustable" (dimensionless) constant ([24], p. 233).

If, moreover, we limit our focus to kink-type waveforms, as physical intuition suggests, and have the piston located at $x = -\infty$ and moving to the right along the x-axis, then $u_x < 0$ and Equation (32) becomes

$$\mu = -b_1^2 r^2 \varrho \phi_{xx}, \tag{34}$$

where we have used the relation $u = \phi_x$. Returning to our dimensionless variables, and once again applying the finite-amplitude approximation, it is readily established that, under the aforementioned assumptions, the following (simpler) weakly-nonlinear PDE replaces Equation (5) as our bi-directional EoM:

$$\phi_{tt} - [1 - 2\epsilon(\beta-1)\phi_t]\phi_{xx} + \epsilon(1-\delta_1)\partial_t(\phi_x)^2 = \sigma\phi_{txx} + a_1^2\phi_{ttxx} \qquad \text{(artificial } \mu\text{)}. \tag{35}$$

In Equation (35), which we observe applies *only* to the RRG case, we have set $a_1 := b_1 r\sqrt{2}/L$, where we require that $a_1 \sim \mathcal{O}(\epsilon)$, and

$$\delta_1 := \frac{\epsilon b_1^2 45(1-\chi)^2}{\chi^2} \qquad (0 < \delta_1 < 1), \tag{36}$$

where the requirement $\delta_1 \in (0,1)$ implies that b_1 must satisfy the inequality

$$0 < b_1 < \frac{1}{3\sqrt{5\epsilon}}\left(\frac{\chi}{1-\chi}\right). \tag{37}$$

Assuming the gas at $x = +\infty$ is in its equilibrium state, and thus motionless, and observing that in the present context U_p is the *dimensional* speed of the piston, we let $\phi(x,t) = G(\eta)$, where $\eta = x - v_1 t$ and the (dimensionless) shock speed v_1 is taken to be a positive constant, and then substitute into Equation (35). On integrating once with respect to η and then imposing/enforcing the asymptotic conditions $g \to 1, 0$ as $\eta \to \mp\infty$, respectively, Equation (35) is reduced to the ODE

$$a_1^2 v_1 g'' - \sigma g' - \epsilon\beta_1 g(1-g) = 0, \tag{38}$$

where we note that the resulting constant of integration is zero. In Equation (38), $g(\eta) = G'(\eta)$, where a prime denotes $\mathrm{d}/\mathrm{d}\eta$; we have defined

$$\beta_1 := \beta - \delta_1, \tag{39}$$

recalling that β is the coefficient of nonlinearity (see Equation (6)); and

$$v_1 = \tfrac{1}{2}\epsilon\beta_1 + \sqrt{1 + \tfrac{1}{4}\epsilon^2\beta_1^2} \qquad (v_1 > 1), \tag{40}$$

which we observe is the positive root of

$$v_1^2 - \epsilon\beta_1 v_1 - 1 = 0. \tag{41}$$

To apply the solution methodology employed in (Ref. [2], §2) to Equation (38), the following condition must be satisfied:

$$25a_1^2(v_1^2 - 1) = 6\sigma^2. \tag{42}$$

In (Ref. [2], §2), satisfying Equation (42) required that the value of the Mach number be fixed, a constraint which clearly limits the usefulness of the TWSs presented in that article. Here, however, we shall use this restriction to our advantage; specifically, in the following sense: Since the value of $\tilde{\mu}$ for a given poroacoustic flow is, in general, not known, and we are seeking a kink-type TWS, then the only possible solution of Equation (42) in the present context is

$$\sigma = 5a_1\sqrt{\frac{v_1^2 - 1}{6}} \qquad \Longrightarrow \qquad \tilde{\mu} = \frac{5\varrho_0 b_1 r}{\chi}\sqrt{\frac{\mathfrak{v}_1^2 - c_0^2}{3}}, \tag{43}$$

where we observe that $\mathfrak{v}_1 = v_1 c_0$ is the *dimensional* shock speed and, moreover, that $v_1 > 1$ implies $\mathfrak{v}_1 > c_0$.

On imposing the usual wave front condition $g(0) = 1/2$, but otherwise referring the reader to (Ref. [2], §2) for details regarding its derivation, the TWS we seek is

$$g(\eta) = \frac{1}{4}\mathrm{sech}^2\left(\frac{27\eta}{16\lambda_1} + \mathcal{K}\right) + \frac{1}{2}\left[1 - \tanh\left(\frac{27\eta}{16\lambda_1} + \mathcal{K}\right)\right], \tag{44}$$

where $\mathcal{K} = \tanh^{-1}\left(-1 + \sqrt{2}\right)$. Letting $\lambda_1 = \ell_1/L$ denote the *dimensionless* shock thickness (Recall Equation (2)) admitted by Equation (44), it is easily established that

$$\lambda_1 = \frac{135a_1^2 v_1}{8\sigma} = \frac{81a_1 v_1}{4\sqrt{6(v_1^2 - 1)}} \qquad \Longrightarrow \qquad \ell_1 = \frac{81\mathfrak{v}_1 b_1 r}{4\sqrt{3(\mathfrak{v}_1^2 - c_0^2)}}, \tag{45}$$

where ℓ_1 is the corresponding *dimensional* shock thickness. Also, with regard to computing λ_1, it should be noted that $g''(\eta^*) = 0$, where $\eta^*(< 0)$ is given by

$$\eta^* = \frac{10a_1^2 v_1\left[\tanh^{-1}(1/3) - \mathcal{K}\right]}{\sigma}, \tag{46}$$

and where it should also be noted that $g(\eta^*) = 5/9$.

The usefulness of Equation (45) might be ascertained as follows. Assume that $\mathfrak{v}_1$ and ℓ_1 can both be determined, either directly or indirectly, from experimental measurements and, moreover, that both are (at most) slowly varying functions of time. With $\mathfrak{v}_1$ known, b_1 can, of course, be computed using Equations (36), (39) and (40). If this (inferred) value of b_1 satisfies the inequality in Equation (37), $a_1 \sim \mathcal{O}(\epsilon)$ is also satisfied, and the measured value of ℓ_1 is in agreement with that computed from

Equation (45) over, say, some span of time $t \in \mathcal{T}$, then we can expect Equation (45) to prove useful as an approximation within the transition region of our kink-type traveling wave profile for $t \in \mathcal{T}$.

6. Discussion: Possible Follow-On Studies

In addition to gaining a better understanding of how the solution of IBVP (30) behaves for large-T, in particular, determining to what extent (if any) the recurrence behavior seen in Figures 2c and 3c is related to the functional form of a given IC, future work on poroacoustic RRG theory could included the use of homogenization methods in problems wherein K and/or χ vary with position. Other possible extensions include the poroacoustic generalization of the study carried out in Ref. [25], wherein α was taken to be a function of $(u_x)^2$, and also the case in which $\tilde{\mu}$ is a power-law function of the shear rate tensor. Follow-on work might also include the study of poroacoustic signaling problems involving sinusoidal and/or shock input signals, as well as problems in which changes in entropy and temperature are taken into account.

Funding: This work was supported by U.S. Office of Naval Research (ONR) funding.

Acknowledgments: The author thanks Ivan Christov, for kindly providing him with the specialized MATHEMATICA code used to generate the data sets from which Figures 1–3 were plotted, and Markus Scholle, for the kind invitation to contribute to this special issue. The author also thanks the two anonymous reviewers for the critiques they provided; their constructive comments have led to a marked improvement in this paper.

Conflicts of Interest: The author declares no conflict(s) of interest.

Appendix A. Corrections to Ref. [13]

As kindly confirmed by Prof. Leibovich [26], Equations (4b), (4d), and (13), and the caption of FIG. 2, in Ref. [13] contain typographical errors; the required corrections, also provided by Prof. Leibovich [26], read as follow:

- In Equation (4b), replace a_1 with a_0.
- In Equation (4d), replace the exponent $3/2$ with $-3/2$.
- In Equation (13), delete the factor ν.
- In the caption of FIG. 2, replace (16) with (15).

References

1. Rubin, M.B.; Rosenau, P.; Gottlieb, O. Continuum model of dispersion caused by an inherent material characteristic length. *J. Appl. Phys.* **1995**, *77*, 4054–4063. [CrossRef]
2. Jordan, P.M.; Keiffer, R.S.; Saccomandi G. A re-examination of weakly-nonlinear acoustic traveling waves in thermoviscous fluids under Rubin–Rosenau–Gottlieb theory. *Wave Motion* **2018**, *76*, 1–8. [CrossRef]
3. Burmeister, L.C. *Convective Heat Transfer*, 2nd ed.; Wiley: Hoboken, NJ, USA, 1993; §2.8.
4. Nield, D.A.; Bejan, A. *Convection in Porous Media*, 5th ed.; Springer: Berlin/Heidelberg, Germany, 2017; Chapter 1.
5. Straughan, B. *Stability and Wave Motion in Porous Media*; Applied Mathematical Sciences; Springer: Berlin/Heidelberg, Germany, 2008; Volume 165.
6. Morduchow, M.; Libby, P.A. On a complete solution of the one-dimensional flow equations of a viscous, heat-conducting, compressible gas. *J. Aeronaut. Sci.* **1949**, *16*, 674–684, 704. [CrossRef]
7. Thompson, P.A. *Compressible-Fluid Dynamics*; McGraw–Hill: New York, NY, USA, 1972.
8. Keiffer, R.S.; McNorton, R.; Jordan, P.M.; Christov, I.C. Dissipative acoustic solitons under a weakly-nonlinear, Lagrangian-averaged Euler-α model of single-phase lossless fluids. *Wave Motion* **2011**, *48*, 782–790. [CrossRef]
9. Beyer, R.T. The parameter B/A. In *Nonlinear Acoustics*; Hamilton, M.F., Blackstock, D.T., Eds.; Academic Press: Cambridge, MA, USA, 1998; pp. 25–39.
10. Crighton, D.G. Model equations of nonlinear acoustics. *Ann. Rev. Fluid Mech.* **1979**, *11*, 11–33. [CrossRef]
11. Ott, E.; Sudan, R.N. Damping of solitary waves. *Phys. Fluids* **1970**, *13*, 1432–1434. [CrossRef]
12. Karpman, V.I. Soliton evolution in the presence of perturbation. *Phys. Scr.* **1979**, *20*, 462–478. [CrossRef]
13. Leibovich, S.; Randall, J.D. On soliton amplification. *Phys. Fluids* **1979**, *22*, 2289–2295. [CrossRef]

14. Zabusky, N.J.; Kruskal, M.D. Interaction of "Solitons" in a collisionless plasma and the recurrence of initial states. *Phys. Rev. Lett.* **1965**, *15*, 240–243. [CrossRef]
15. Rossmanith D.; Puri A. Recasting a Brinkman-based acoustic model as the damped Burgers equation. *Evol. Equs. Control Theory* **2016**, *5*, 463–474. [CrossRef]
16. Nimmo, J.J.; Crighton, D.G. Bäcklund transformations for nonlinear parabolic equations: The general results. *Proc. R. Soc. Lond. A* **1982**, *384*, 381–401.
17. Malfliet, W. Approximate solution of the damped Burgers equation. *J. Phy. A Math. Gen.* **1993**, *26*, L723–L728. [CrossRef]
18. Ciarletta, M.; Straughan, B. Poroacoustic acceleration waves. *Proc. R. Soc. A* **2006**, *462*, 3493–3499. [CrossRef]
19. Jordan, P.M. Growth and decay of acoustic acceleration waves in Darcy-type porous media. *Proc. R. Soc. A* **2005**, *461*, 2749–2766. [CrossRef]
20. Jordan, P.M. Poroacoustic solitary waves under the unidirectional Darcy–Jordan model. *Wave Motion* **2020**, *94*, 102498. [CrossRef]
21. Crighton, D.G. Propagation of finite-amplitude waves in fluids. In *Handbook of Acoustics*; Crocker, M.J., Ed.; Wiley-Interscience: Hoboken, NJ, USA, 1998; Chapter 17.
22. Logan, J.D. *An Introduction to Nonlinear Partial Differential Equations*, 2nd ed.; Wiley-Interscience: Hoboken, NJ, USA, 2008.
23. Von Neumann, J.; Richtmyer, R.D. A method for the numerical calculation of hydrodynamic shocks. *J. Appl. Phys.* **1950**, *21*, 232–237. [CrossRef]
24. Roache, P.J. *Computational Fluid Dynamics*; Hermosa Publishers: Albuquerque, NM, USA, 1972.
25. Jordan, P.M.; Saccomandi, G. Compact acoustic travelling waves in a class of fluids with nonlinear material dispersion. *Proc. R. Soc. A* **2012**, *468*, 3441–3457. [CrossRef]
26. Leibovich, S. (Cornell University, New York, NY, USA). Personal communication; 22 May 2018.

Article

Study on the Characteristic Point Location of Depth Average Velocity in Smooth Open Channels: Applied to Channels with Flat or Concave Boundaries

Tongshu Li [1], Jian Chen [2], Yu Han [1,*], Zhuangzhuang Ma [1] and Jingjing Wu [1]

1 College of Water Resources & Civil Engineering, China Agricultural University, 17 Tsinghua East Rd., Haidian District, Beijing 100083, China; cau_litongshu@163.com (T.L.); caumjf@163.com (Z.M.); s20193091621@cau.edu.cn (J.W.)
2 College of Engineering, China Agricultural University, 17 Tsinghua East Rd., Haidian District, Beijing 100083, China; jchen@cau.edu.cn
* Correspondence: yhan@cau.edu.cn; Tel.: +86-182-0161-0386

Received: 14 November 2019; Accepted: 2 February 2020; Published: 5 February 2020

Abstract: Based on the flow partition theory, we derive a mathematical expression by using the log-law for the characteristic point location (CPL) of depth average velocity in channels with flat or concave boundaries. It can manifest the position of the characteristic points in the vertical direction relative to the channel side wall or bed. Taking rectangular and semi-circular channels as research objects, we put forward a method to calculate the discharge of channels with CPL. Additionally, we carried out some experiments on rectangular and semi-circular channel sections. CPL's analytic expression is validated against experimental results through comparison of velocity and discharge. The proposed formulation of characteristic point location could be extensively employed in flow measurements of flat and concave boundary channels, which has practical application value in simplifying the flow measurement steps of open channels.

Keywords: log-law; flow partitioning theory; characteristic point location; velocity; discharge

---

1. Introduction

The contradiction between the serious shortage of agricultural water resources and unreasonable utilization has been restricting the development of agricultural economy [1]. Precision agriculture provides the potential to enhance irrigation efficiency through specific methods of flow measurement in channels.

Flow measurement methods through weir-gate structures [2] and velocity–area methods are widely used in flow measurement. The principle of a sluice gate or weir is based on the concept of critical flow, making it possible to measure only the depth and calculate the flow rate, thus simplifying the continuous monitoring of flow rate [3]. However, critical flow in open channels can be formed through two general methods: raising the bottom of the channel or contracting the cross-sectional area of the flow [3,4]. Neither of these two methods is easy to implement. Impurities on the sluice gate or weir accumulate easily and need to be cleaned regularly, which is difficult to achieve during actual production. Apart from measuring with weir-gate structures, the velocity–area method for calculating discharge in an open channel is considered to be particularly reliable. However, a large number of measurements results in an associated time cost of measurement. Numerous researchers have proposed different forms of the velocity distribution law in order to predict flow discharge accurately.

The study of velocity distribution is critical to the calculation of the discharge. Relevant research can be traced back to the 1930s. In 1938, Keulegen [5] first proposed that the log-law could be used to describe the time-averaged velocity distribution of fully developed open channel uniform turbulence:

$$\frac{u}{u_*} = \frac{1}{\kappa} \ln \frac{u_* \times y}{\nu} + \omega \quad (1)$$

where u is the velocity in the channel, u_* is frictional velocity, ν is the coefficient of kinematic viscosity, κ is the von Kármán constant, and ω is constant.

Nikuradse made an artificial sand rough pipe using pasting sand with a uniform particle size on the wall of a pipe [6,7]. The analysis of the experimental data of the smooth sand tube showed that the values of parameters κ and ω in log-law are 0.4 and 5.5. Different researchers have given different values of the von Kármán constant κ and constant ω according to their experimental data [8–10]. This paper uses the parameter values given by Nikuradse for further derivation.

Under the guide of velocity distribution, researchers have used different methods to measure velocity. Chiu [11–13] developed the entropy model, correlating the mean flow velocity to the maximum value using the linear relationship $Um = \varphi(M)U_{max}$, which depends on the entropic parameter M. Maghrebi [14] used the similarity between the magnetic field of a current wire and the isovel contours in a channel cross-section to derive the isovel patterns. Then, one can easily obtain the discharge using a single point on the isovel patterns of velocity measurement. However, the measured points are only selected from the upper half of the water depth and away from the boundaries, which would be much closer to the measured discharge. Bretheim et al. [15] used the restricted nonlinear (RNL) model to simulate wall turbulence and obtained the real mean velocity distribution in the channel at a low Reynolds number. Hong et al. [16] employed a systematic measuring technology combining ground-penetrating radar and surface velocity radar and established the rating curves representing the relation of water surface velocity to the channel cross-sectional mean velocity and flow area. Then, stream discharge was deduced from the resulting mean velocity and flow area. Moramarco et al. [17] proposed a new method for estimating discharge from surface velocity monitoring (u_{surf}). Based on entropy theory and sampling u_{surf}, it can identify the two-dimensional velocity distribution in the cross-sectional flow area. This method is more accurate for rivers with a lower aspect ratio where secondary currents are expected. Johnson and Cowen [18,19] predicted the mean streamwise velocity and the depth-averaged velocity by permitting remote determination of the velocity power-law exponent. Then, the volumetric discharge from surface measurements of currents can be determined. Khuntia et al. [20] presented a new methodology to predict the depth-averaged velocity. They used multi-variable regression analysis to develop five models to predict the point velocities in terms of non-dimensional geometric and flow parameters at any desired location. Through these attempts, researchers have discussed new thoughts of quantifying channel discharge. But few people determine the average velocity of a channel by looking for the location of characteristic points.

Overall, this study investigates a new concept of characteristic point location aimed at estimating the discharge in open channels, which is based on flow partitioning theory and the log-law. This method is applied to flat channels (e.g., a rectangular channel) and concave channels (e.g., a semi-circle channel). The derived theoretical expression of characteristic point location is also verified by the experimental data. We expect that the method for flow measurement conveniently proposed by this paper could be used widely for different varieties of irrigation channels.

2. Methodology

This paper assumes that characteristic points exist on the channel cross-section whose velocities could be called characteristic velocities. We can measure these points to get the precise measurements of velocity and discharge. The positions of these points are called characteristic point locations of depth average velocity (CPL).

2.1. CPL in Rectangular Channels

2.1.1. Existing Form of the Division Line in Rectangular Channels

Consider the case of steady, uniform flow in a smooth rectangular channel with an aspect ratio of b/h, where h = water depth and b = channel width. Due to existence of a secondary current, the division line is actually a zero Reynolds shear stress line in the flow area, and also a line without energy cross-transfer [21]. Yang et al. [22] have proven the physical existence of division lines by using experimental data recorded at the Hydraulics Laboratory, University of Wollongong, as well as the experimental data recorded previously by Melling and Whitelaw [23] and that of Tracy [24].

Einstein [25] proposed that the flow cross-sectional area can be divided into three sub-sections that correspond to the side wall and channel bed. However, he did not propose any method of determining the exact location of the division line. Chien and Wang [26] performed an in-depth study of Einstein's idea. They gave the total overall boundary shear stress, τ_0, as:

$$\tau_0 = \rho g R S_e \tag{2}$$

where ρ = fluid density; g = gravitational acceleration; R = hydraulic radius; and S_e = energy slope.

Then, they divided τ_0 into two parts corresponding to the bed, $\overline{\tau_b}$, and side wall, $\overline{\tau_w}$. They also defined two hydraulic radii, one for the wall R_w, and one for the bed R_b. Thus:

$$\overline{\tau_w} = \rho g R_w S_e = \rho g S_e A_w / h \tag{3}$$

$$\overline{\tau_b} = \rho g R_b S_e = \rho g S_e A_b / b \tag{4}$$

where $\overline{\tau_b}$ = mean bed shear stress;$\overline{\tau_w}$ = mean wall shear stress; A_b = flow area corresponding to channel bed; A_w = flow area corresponding to channel side wall; R_w = hydraulic radius corresponding to side wall; and R_b = hydraulic radius corresponding to bed.

According to Yang [27], the mechanical energy contained in any flow will be transmitted to the boundary nearest to the "relative distance" and dissipated. The dimensionless relative distance Φ is defined as the ratio of geometrical length between a point in the flow field and the characteristic boundary length, representing the energy dissipation capacity of the boundary (for a smooth boundary channel, this would be the viscous layer thickness D_s).

The side wall and bed will each have their own fair share of surplus energy transferred from the main flow depending on the minimum relative distance Φ. Accordingly, for any unit volume in the flow field, there are two possible ways to transfer the surplus energy: either toward the bed (Φ_b) or toward the side wall (Φ_w). If $\Phi_w \leq \Phi_b$, the energy will be transferred toward the side wall. Conversely, if $\Phi_w \geq \Phi_b$, then the energy will be transported to the bed. It follows that the condition where $\Phi_w = \Phi_b$ will define a division line by which the flow region near the corner is divided into two sub-flow sections.

Therefore, the division line ($\Phi_w = \Phi_b$) in Figure 1a can be expressed as:

$$\frac{z}{D_{sw}} = \frac{y}{D_{sb}} \tag{5}$$

where z and y denote the horizontal and vertical axes in the Cartesian coordinates system, respectively; y = vertical direction normal to bed; z = transverse flow direction;D_{sw} = characteristic length for smooth side wall; and D_{sb} = characteristic length for smooth channel bed.

Substituting $D_{sw} = C_s \nu / \overline{u_{*w}}$ and $D_{sb} = C_s \nu / \overline{u_{*b}}$ into Equation (5) gives:

$$z = ky;\ k = \frac{\overline{u_{*b}}}{\overline{u_{*w}}}. \tag{6}$$

where C_s = constant; $\overline{u_{*w}}$ = mean side wall shear velocity; and $\overline{u_{*b}}$ = mean bed shear velocity. The geometric significance of k in Figure 1a is the reciprocal of the slope of the line OD. Point D is the intersection of the division line and the water surface line.

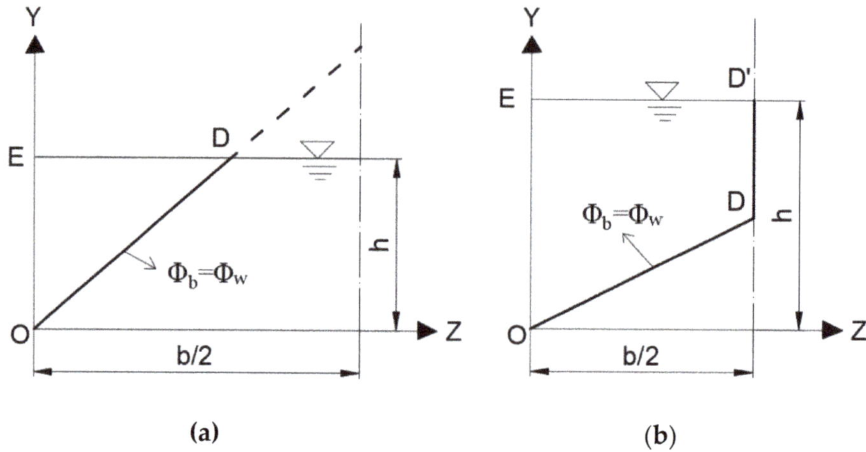

Figure 1. Division lines for smooth rectangular channels: (**a**) $b/h \geq \alpha$; (**b**) $b/h \leq \alpha$.

Then, it can be obtained that:

$$R_w = \frac{A_w}{h} = \frac{\frac{1}{2}kh^2}{h} \tag{7}$$

$$R_b = \frac{A_b}{b} = \frac{bh - kh^2}{b} \tag{8}$$

Using Equations (3)–(8), k can be evaluated, yielding:

$$\frac{\overline{\tau_b}}{\overline{\tau_w}} = \frac{\rho\overline{u_{*b}}^2}{\rho\overline{u_{*w}}^2} = k^2 = \frac{1 - \frac{kh}{b}}{\frac{k}{2}} \tag{9}$$

Or:

$$k^3 + (2h/b)k - 2 = 0; \frac{b}{h} \geq \alpha \tag{10}$$

Using a similar method, Yang and Lim [27] also gave the evaluation of k_1 in Figure 1b:

$${k_1}^3 + (b/2h)k_1 - 2 = 0; \frac{b}{h} \leq \alpha \tag{11}$$

where k_1 represents the slope of the division line OD in Figure 1b.

Highly inspired by the derivation of Yang and Lim [27], we also take the two cases of Figure 1 to analyze. It has been proven that the critical aspect ratio α is equal to 2 in smooth rectangular channels by Yang and Lim [27]. As the rectangle is axisymmetric, we take the vertical line of the channel section as the axis and only analyze half of it.

(a) Determination of the location of division line in wide–shallow channel ($b/h \geq 2$)

The division line can be determined by Equation (10) when $b/h \geq 2$ (Figure 1a). Since $2h/b$ is a constant, solving Equation (10) will give the value of k.

Referring to the value of k, a comparison relationship between $1/k$ and $2h/b$ is as follows:

$$\frac{1}{k} - \frac{2h}{b} \geq 0 \tag{12}$$

It can be shown from Inequality (13), so

$$\frac{1}{k} \geq \frac{2h}{b}; \frac{2h}{b} \leq 1 \tag{13}$$

That is to say, when $b/h \geq 2$, the slope of the division line is always greater than $2h/b$. The intersection point of the division lines is on or above the water surface line.

(b) Determination of the location of division line in narrow–deep channel ($b/h \leq 2$)

The division line can be determined by Equation (11) when $b/h \leq 2$ (Figure 1b). Since $b/2h$ is a constant, the value of k_1 can be obtained by solving Equation (11).

Referring to the value of k_1, a comparison relationship between k_1 and $2h/b$ is as follows:

$$k_1 - \frac{2h}{b} \geq 0 \tag{14}$$

It can be shown from Inequality (15), so

$$k_1 \geq \frac{2h}{b}; \frac{b}{2h} \leq 1 \tag{15}$$

That is to say, when $b/h \leq 2$, the slope of the division line is always greater than $2h/b$. The intersection point of the division lines is on or above the water surface line, the same as $b/h \geq 2$.

By combining cases (**a**) and (**b**), we know that Figure 1b will not occur. Whether in a wide–shallow or in a narrow–deep channel, the existing form of the division line is always shown as Figure 1a.

2.1.2. CPL of Lines in Rectangular Channel

As shown in Figure 2, taking the left half cross-section of the channel as an example, the dividing line divides the cross-section of the channel into regions I (ABCD), II (OAD) and III (ODE).

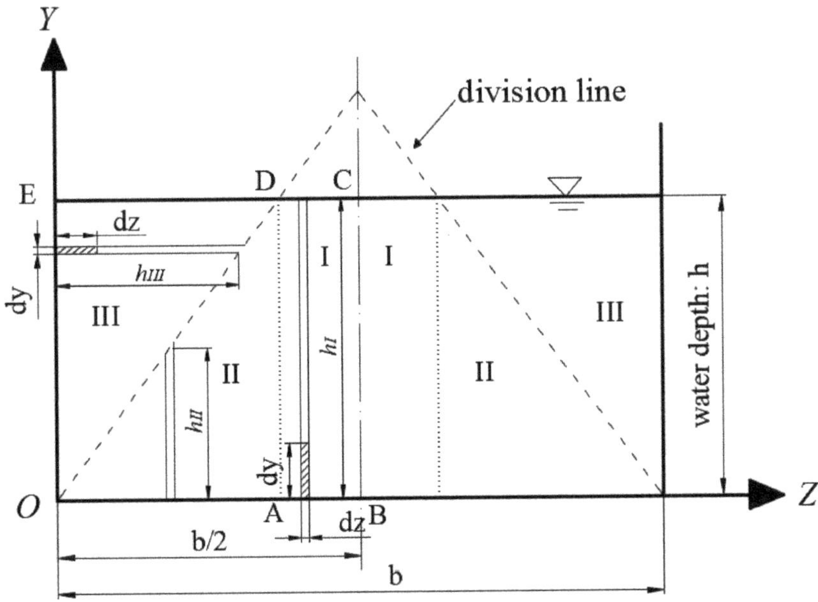

Figure 2. Flow partitioning in rectangular channels.

In region I, the discharge of the shadow rectangle area can be expressed using Equation (16):

$$dQ = udA = udydz \tag{16}$$

where dQ is the discharge of the shadow rectangle area; dA is the area of shadow rectangle; and u is the longitudinal velocity of any point in the area.

By integrating y, the discharge Q can be obtained as follows:

$$Q = dz\int_0^{h_I} udy \tag{17}$$

As the rectangular area can be expressed as $A = h_I dz$, the average velocity of this flow area can be obtained as:

$$\overline{u} = \frac{Q}{A} = \frac{dz\int_0^{h_I} udy}{h_I dz} = \frac{\int_0^{h_I} udy}{h_I} \tag{18}$$

According to the log-law of vertical velocity in channel section:

$$\frac{u}{u_*} = 2.5\ln\left(\frac{u_* y}{\nu}\right) + 5.5 \tag{19}$$

Substituting Equation (19) to (18) gives:

$$\overline{u} = 2.5u_* \ln\frac{u_* \cdot h_I}{\nu} + 3u_* \tag{20}$$

Combining Equations (19) and (20) gives:

$$\overline{u} = 2.5u_* \ln\frac{u_* \cdot h_I}{\nu} + 3u_* = 2.5u_* \ln\frac{u_* \cdot y}{\nu} + 5.5u_* \tag{21}$$

y is expressed as:

$$y = y_I = \frac{h_I}{e} = \frac{h}{e} \tag{22}$$

where y_I represents the location of depth average velocity in region I, and $h_I (= h)$ is the water depth of the channel.

In region II, y_{II} represents the CPL of depth average velocity, which is similar to region I:

$$y_{II} = \frac{h_{II}}{e} \tag{23}$$

where y_{II} represents the location of depth average velocity in region II, and h_{II} is the distance perpendicular to the channel bed to the division line.

In region III, as shown in Figure 2, the discharge of the shadow rectangle area can be expressed as:

$$dQ = udA = udydz \tag{24}$$

Integrating z along the direction of z axis, the discharge can be determined using Equation (25):

$$Q = dy\int_0^{h_{III}} udz \tag{25}$$

where h_{III} is the horizontal distance from the side wall of the channel to the division line.

As the rectangular cross-sectional area can be expressed as $A = h_{III}dy$, the average velocity of this flow area can be obtained as shown as Equation (26):

$$\overline{u} = \frac{Q}{A} = \frac{dy\int_0^{h_{III}} udz}{h_{III}dy} = \frac{\int_0^{h_{III}} udz}{h_{III}} \tag{26}$$

According to the log-law of vertical velocity in channel section:

$$\frac{u}{u_*} = 2.5\ln\left(\frac{u_* z}{\nu}\right) + 5.5 \tag{27}$$

Using Equations (26) and (27), z can be evaluated using:

$$\frac{\frac{\int_0^{h_{III}} udz}{h_{III}}}{u_*} = 2.5\ln\left(\frac{u_* z}{\nu}\right) + 5.5 \tag{28}$$

Giving:

$$z = z_{III} = \frac{h_{III}}{e} \tag{29}$$

where z_{III} represents the location of the depth average velocity in region III.

2.1.3. CPL of Regions in Rectangular Channel

For region I, according to Equation (22), since the water depth h in the channel is constant, measuring the velocity of the CPL of depth average velocity can provide the average velocity for the whole section of region I. Therefore, the average discharge can be obtained by multiplying the average cross-section velocity by the area of the region. Similarly, the flow in regions II and III can be obtained using the same method. Figure 3 shows that P_I, P_{II}, and P_{III} are the characteristic points that represent the mean velocities in regions I, II, and III, respectively.

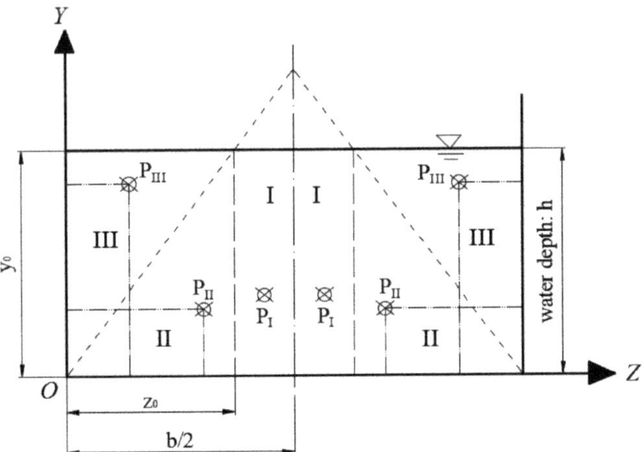

Figure 3. Location map of the characteristic points in different regions.

For region II, the discharge Q can be expressed using Equation (30):

$$Q = u_* \int_0^{z_0} \frac{z}{k}\left(2.5\ln\left(\frac{\sqrt{gS}\left(\frac{z}{k}\right)^{1.5}}{e\nu}\right) + 5.5\right)dz = \frac{u_*}{k}\left(z_0{}^2\left(1.25\ln\left(\frac{\sqrt{gS}}{e\nu}\left(\frac{z_0}{k}\right)^{1.5}\right) + 1.8125\right)\right) \tag{30}$$

where e is a constant; S is the slope of channel; and z_0 is shown in Figure 3.

The mean velocity can be calculated from the value of Q/A, which can be expressed using Equation (31):

$$\overline{u} = \frac{Q}{A} = \frac{2\frac{u_*}{k}\left(z_0{}^2\left(1.25\ln\left(\frac{\sqrt{gS}}{ev}\left(\frac{z_0}{k}\right)^{1.5}\right) + 1.8125\right)\right)}{z_0 h_0} \tag{31}$$

Using Equations (6), (19) and (31), Equation (32) can be obtained:

$$u_*\left(2.5\ln\left(\frac{\sqrt{gS}\left(\frac{z_0}{k}\right)^{1.5}}{ev}\right) + 3.625\right) = u_*\left(2.5\ln\left(\frac{\sqrt{gS}\left(\frac{z}{k}\right)^{1.5}}{ev}\right) + 5.5\right) \tag{32}$$

Therefore, the y and z value of the point P_{II} can be calculated using Equation (33):

$$\begin{cases} \overline{z_{II}} = \frac{z_0}{e^{0.125}} \\ \overline{y_{II}} = \frac{y_0}{e^{1.125}} \end{cases} \tag{33}$$

where $\overline{z_{II}}$ and $\overline{y_{II}}$ represent the coordinates of P_{II}.

For region III, the analysis method is similar to that of region II. The z and y value of the point P_{III}, which can be used to calculate the value of Q of region III, can be expressed using Equation (34):

$$\begin{cases} \overline{z_{III}} = \frac{z_0}{e^{1.125}} \\ \overline{y_{III}} = \frac{y_0}{e^{0.125}} \end{cases} \tag{34}$$

where $\overline{z_{III}}$ and $\overline{y_{III}}$ represent the coordinates of P_{III}.

2.2. CPL in Semi-Circular Channel

The concave boundary cross-section is one of the most commonly used cross-section forms in farmland canal systems, open-flow culverts, urban drainage pipelines and other projects. Compared with the rectangular section, it has a larger hydraulic radius under the same conditions, so the velocity of the cross section is larger and the water flow that can be transported is also larger [28]. In order to obtain the discharge of the concave boundary cross-section channel more quickly and accurately, it is necessary to master the velocity distribution of the cross-section. Figure 4 shows the semi-circle channel cross-section.

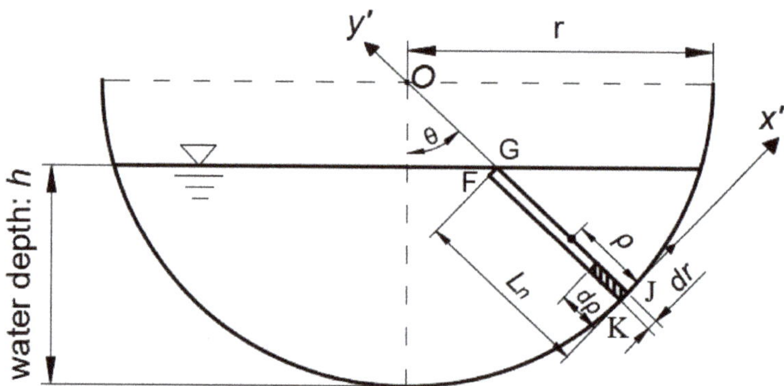

Figure 4. Section diagram of semi-circular channel.

As shown in Figure 4, a straight line, which crosses the center of the circle and is angled with the center line (θ), is taken from a semicircular channel with radius R and maximum water depth h.

The coordinate system is established by taking the straight line and the direction perpendicular to the straight line as the coordinate axis. According to Figure 4, the rectangle FGJK is L_n in length and dr in width, where Ln is the length of the underwater part of the radius. With the rectangle FGJK, the area is $A_{FGJK} = L_n \cdot dr$ and the discharge through FGJK can be expressed as:

$$Q = u \cdot A_{FGJK} = u \cdot L_n \cdot dr \tag{35}$$

Then, taking a shadow rectangle whose edge lengths are $d\rho$ and dr ($d\rho$ and dr are infinitesimal), the discharge through the shadow rectangle is:

$$dQ = u \cdot dr \cdot d\rho \tag{36}$$

By integrating $d\rho$ in the direction of y' and substituting the log-law, we get:

$$Q = dr \cdot \int_0^{L_n} u d\rho = dr \cdot u_* \cdot L_n \cdot \left(2.5 \ln \frac{L_n \cdot u_*}{\nu} + 3\right) \tag{37}$$

Substituting the log-law into Equation (35) gives:

$$Q = L_n \cdot dr \cdot u_* \cdot \left(2.5 \ln \frac{u_* \cdot \rho}{\nu} + 5.5\right) \tag{38}$$

For Equations (37) and (38), the coordinates of the characteristic points representing the average velocity of the straight line can be obtained from Equation (39):

$$\rho = \frac{Ln}{e} \tag{39}$$

where ρ is the CPL of the depth average velocity of the line.

2.3. *Discharge of Rectangular and Semi-Circular Channel*

2.3.1. Discharge in a Rectangular Channel

According to Section 2.1.3, based on the CPLs of different regions, the discharge from the cross-section can be given by Equation (40):

$$Q_{rec} = 2 \times (A_I u_{P_I} + A_{II} u_{P_{II}} + A_{III} u_{P_{III}}) \tag{40}$$

where A_I, A_{II} and A_{III} represent the areas of three regions, and u_{P_I}, $u_{P_{II}}$ and $u_{P_{III}}$ represent the velocities of points P_I, P_{II}, and P_{III}.

2.3.2. Discharge in a Semi-Circular Channel

Similar to a rectangular channel, in a semi-circle channel, after obtaining the CPL of every normal line, it is necessary to know how to calculate the discharge through simplified point distribution. As shown in Figure 5, two straight lines passing through the center of the circle have divided the angle into $\theta 1$, $\theta 2$ and $\theta 3$. The discharge can be expressed as Equation (41):

$$Q_{cir} = 2 \times (\alpha_c \cdot A_1 \cdot u_{CPL_1} + A_2 \cdot \frac{u_{CPL_1} + u_{CPL_2}}{2} + \alpha_c \cdot A_3 \cdot u_{CPL_2}) \tag{41}$$

where A_1, A_2 and A_3 represent sub-areas, u_{CPL} is the velocity of the normal line and α_c is the velocity coefficient along the bank. α_c can be determined according to the test conditions.

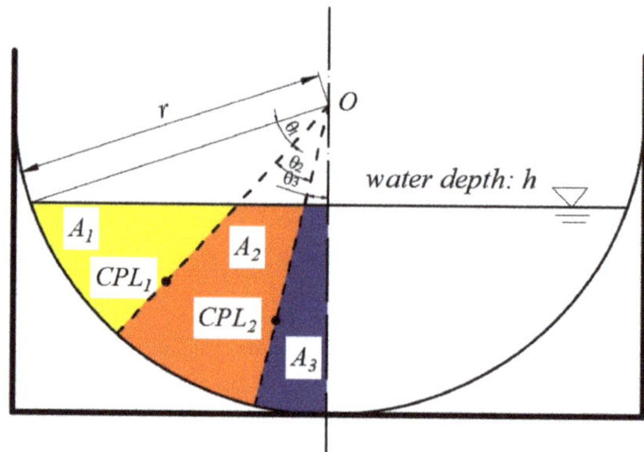

Figure 5. Discharge calculation in a semi-circular channel with the two-line method.

3. Experimental Setup

Table 1 summarizes the test conditions and relevant parameters of a channel cross-section. The rectangle and semi-circle channel section tests are shown in the table. In order to increase the reliability of the test data, the tests are carried out in different places, such as the Fluid Laboratory of the University of Wollongong, the Hydraulic Experiment Hall of the China Agricultural University and the Semi-Circular Test Open Channel of the British Council for Scientific and Engineering Research [29]. Different groups were set up by changing the discharge (or water depth) of the channel. Using acoustic doppler velocimetry (ADV) and other velocity measurement tools to measure the vertical velocity of a cross-section, a large amount of velocity data can be obtained.

Table 1. Summary of experimental conditions.

Cross-Sectional Shape	Conditions	Channel Width: b (m)	Channel Radius: r (m)	Flow Discharge: Q (m^3/s)	Water Depth: h (m)
Rectangle	R1 [a]	0.3	-	0.004	0.065
	R2 [a]			0.004	0.091
	R3 [a]			0.004	0.110
	R4 [b]	0.8	-	0.031	0.120
	R5 [b]			0.033	0.128
	R6 [b]			0.039	0.137
	R7 [b]			0.044	0.153
Semi-circle	S1 [c]	-	0.120	0.005	0.0813
	S2 [c]			0.012	0.1200
	S3 [a]	-	0.150	0.003	0.075
	S4 [a]			0.005	0.085
	S5 [a]			0.008	0.100

[a] Experiment data are from the Fluid Laboratory of the University of Wollongong(UOW). [b] Experiment data are from the Hydraulic Experiment Hall of China Agricultural University(CAU). [c] Experiment data are from Knight [29].

4. Results and Discussion

4.1. Analysis with Rectangular Channels

The consistency of CPL of R1, R2 and R3 in rectangular channels is analyzed in Figure 6. These scatter points cover three regions of the rectangular cross-section. The abscissa of the scatter points in the figure represents the theoretical value of CPL and is calculated by Equations (22), (23) and (29). The longitudinal coordinates of points represent the measured value of CPL, which can be obtained by interpolating the experimental data. It can be seen that the data points are mostly distributed within the error line of ±15%. Therefore, it can be proven that the formulas of CPL in Section 2.1.2 can be applied to rectangular cross-section channels.

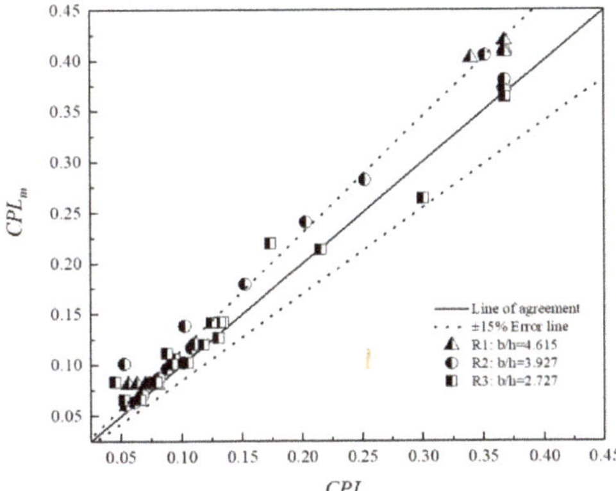

Figure 6. Comparison of the CPL between the theoretical and measured values in a rectangular channel.

In order to further verify the correctness of the theoretical derivation, a series of experiments were conducted at different water depths at the CAU (China). We collated the experimental data of R4-R7 and selected the representative data points to draw in Figure 7. Since the perpendicular bisector is in region I, we calculated the point of CPL with Equation (22). Since the number of measuring points was less, and there were some test errors, the relative error between *CPLt* and *CPLm* still existed. With the increase in water depth, the influence of several measuring points on the results is apparent (R7). The velocities of *CPLt* and *CPLm* have a greater consistency when the number of measuring points is high. Hence, we have reason to think that the velocity of *CPLt* can represent the average velocity of the centerline in rectangular channels.

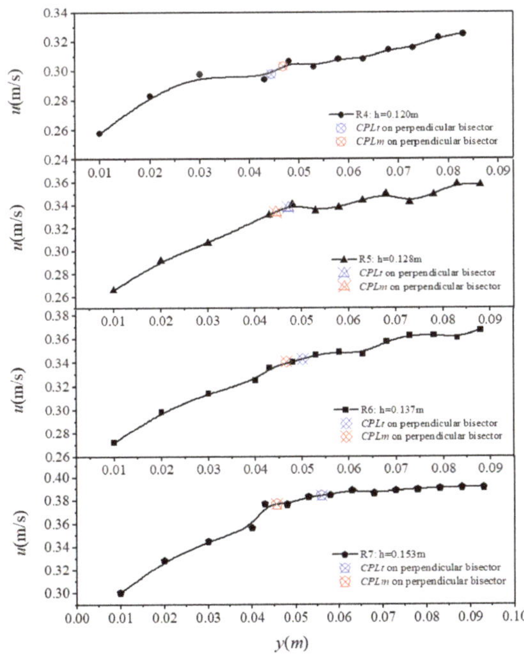

Figure 7. Velocity on the perpendicular bisector with R4-R7 in a rectangular channel.

The characteristic points of depth average velocity in each region are plotted in Figure 8 and recorded as PI, PII and PIII, which are calculated by Equations (22), (33) and (34), respectively. We can obtain the discharge by using the values of PI, PII and PIII in Equation (40). Curves in the graph are isovel patterns in the rectangular cross-section channel based on UOW test data, which is consistent with the contour line measured by Chiu and Chiou [30]. It can be seen from the figures that the velocity near the wall is small and the velocity far from the wall is large. This can be explained by the log-law of velocity distribution. Increasing the monotone leads to the velocity increasing as the distance from the channel side wall or bed increases. Therefore, the maximum velocity point of the cross-section should theoretically be located at the intersection of the vertical line of the channel and the water surface.

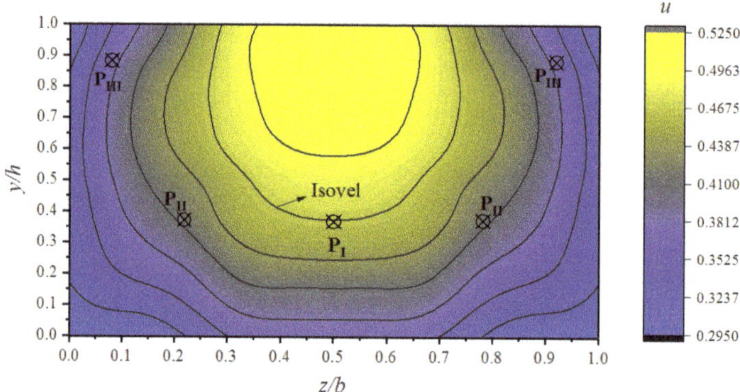

Figure 8. Isovel patterns in a rectangular channel (e.g. R1: b/h = 4.615).

However, in Figure 8, we can see that the maximum velocity point is below the water surface, not on it. This phenomenon was first observed about a century ago [31,32], and further experimental studies showed that it was induced by the presence of secondary cross-sectional flow structures [33]. Due to the occurrence of anisotropic turbulence and cross-sectional secondary currents, which tend to shift the maximum velocity from the free surface to the bed, its identification is still a complex task in hydraulics [34].

4.2. Analysis with Semi-Circular Channels

Mechanical energy is always transmitted to the closest boundary. As shown in Figure 9, this refers to the normal line direction. So, in a concave boundary channel, the analysis is carried out along the normal direction perpendicular to the tangent direction of the wall.

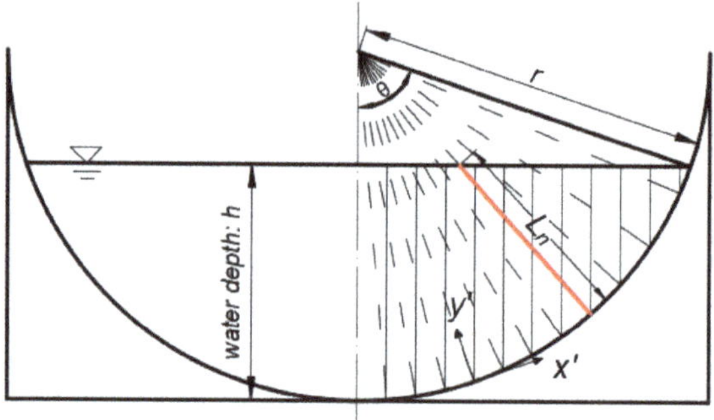

Figure 9. Semi-circular channel cross-section and notation.

In order to compare the accuracy of velocity distribution analysis along the vertical direction and along the normal wall direction, the following analysis has been made. The error values obtained by the experimental condition S2 (in Table 1) and log-law formulas are compared in tabular analysis in Table 2. The calculation formula of the average error value is given by:

$$E = \frac{E_m - E_c}{E_m} \times 100\% \tag{42}$$

where E is the average error value in Table 2, E_m denotes the measured value, and E_c denotes the calculated value.

Table 2. Accuracy comparison of velocity distribution along the vertical wall and normal directions with S2: h = 0.1200.

Average Error Value along Normal Direction		Average Error Value along Vertical Direction	
Normal Slope k_n	Average Error Value E (%)	The Distance from the Vertical Line to the Central Line z/b	Average Error Value E (%)
11.9	2.156661	0.083	3.528655
5.91	3.117373	0.167	3.441878
3.87	2.578154	0.250	2.827718
2.82	3.073467	0.333	7.846099
2.18	3.217222	0.417	6.557529
1.73	2.584048	0.500	10.55338
1.39	2.047317	0.583	12.39980
1.11	2.207014	0.667	19.05022
0.88	2.758453	0.750	21.84237
0.66	4.707039	0.833	30.18244

From Table 2, we can conclude that the average error value obtained by normal analysis along the wall is much lower than that obtained by vertical analysis when consistency analysis of the velocity measurements and the log-law is carried out. Hence, the correctness of the analysis of the concave boundary channel along the normal direction perpendicular to the wall is proved.

Figure 10 shows the comparison of average velocity along the normal line under S1–S5. The horizontal ordinates in the figure are the average value of the velocity at each normal measuring line. The vertical coordinates are obtained by theoretical calculation. Specifically, the CPL is calculated by the formula from Section 2.2, and then the average velocity on the normal line is obtained by substituting it into the log-law.

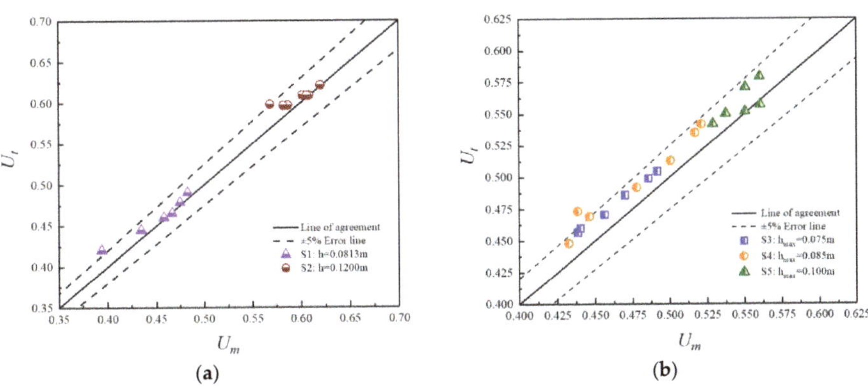

Figure 10. Comparison of measured and theoretical values of mean radial velocity (**a**) S1, S2; (**b**) S3, S4, S5.

Despite some fluctuations of a few test points in Figure 10, most of them are within the error range of 5%. The statistical rules of a large number of test points can still prove the accuracy of the theoretical deduction in Section 2.2. Therefore, the velocity corresponding to the CPL calculated by Equation (39) can represent the average velocity of the channel wall along the normal line to the water surface line. Most points, especially the measuring points close to the water surface, are located above the agreement line, which implies that U_t is bigger than U_m. This is because the log-law appears to deviate near the water surface. Based on this, the log-wake law was proposed by Coles [35], which appears to be the most reasonable extension of the log-law. However, the value of Π in the log-wake law seems not to be universal [36]. So, more research is needed to correct the velocity near the water surface.

As shown in Figure 11, the theoretical and measured values of CPL on each normal line are very close, located in the middle and lower part of the normal line. The solid and dotted lines are smooth arc curves, which are consistent with the concave boundary. Therefore, the accuracy of the formula in Section 2.2 can be proved. It is advisable to use the velocity of CPL to represent the average velocity of the normal line.

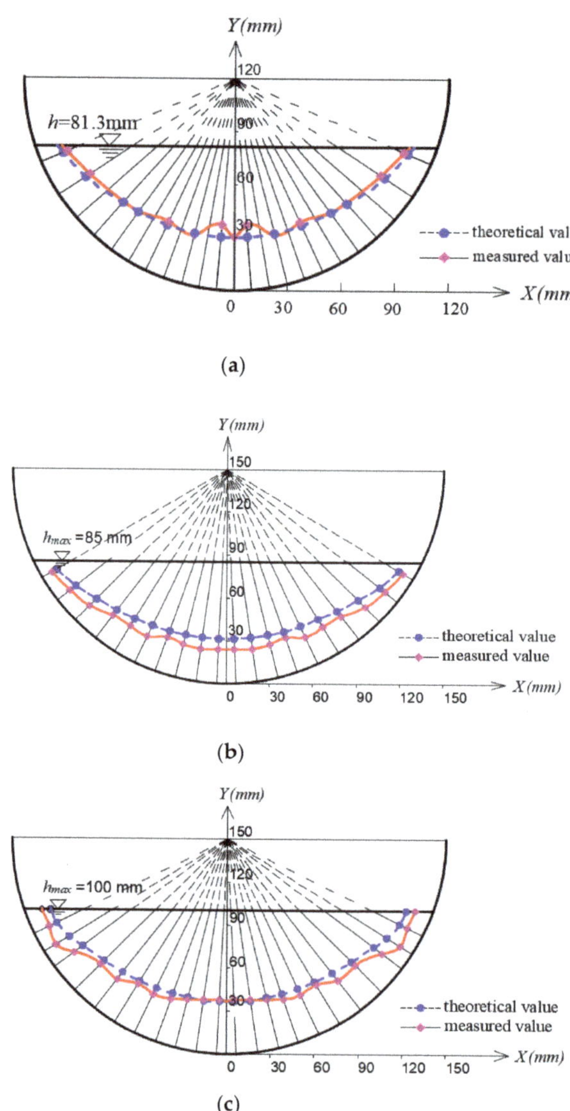

Figure 11. Comparison of measured and theoretical values of CPL (**a**) S1; (**b**) S4; (**c**) S5.

When the S1 condition is applied to Equation (41), we can calculate the discharge of the semi-circular channel. Here, $\alpha_c = 0.8$. Table 3 shows the calculated discharge of the channel with the combination of six kinds of angles. Only the first is a uniform partition; the others are all non-uniform. Relative error refers to the error value of the calculated discharge relative to the measured discharge. It can be found that the relative error value is the smallest only when the angle is divided uniformly, about 5%. Therefore, we can reasonably conclude that the accuracy of the discharge calculation is the highest only if the angle is divided uniformly while using the two-line method.

Table 3. Analysis of calculated and measured discharge under different combinations of angles (S1).

Degree of Angle with the Two-Line Method			**Q Calculation (m^3/s)**	**Q Measurement (m^3/s)**	**Relative Error (%)**
θ_1 (°)	θ_2 (°)	θ_3 (°)			
71/3	71/3	71/3	0.00526	0.005	5.10
18	34	19	0.00565		12.90
18	23	30	0.00531		6.27
29	32	10	0.00551		10.09
35	26	10	0.00542		8.48
36	21	14	0.00533		6.61

5. Summary

This study has investigated a fast flow measurement method based on the distribution of measuring points by conducting laboratory experiments. The principal achievements from the comparison of the theoretical derivation and experimental study are summarized below.

(1) Based on Yang et al.' s partitioning theory [21,22,27], this paper gives a re-description of the existing form of the division line of a rectangular cross-section channel. That is, whether the channel cross-section is wide–shallow or narrow–deep with the center line of the cross-section as the symmetrical axis, and whether the intersection points of the left and right division lines intersect on or above the water surface.

(2) This paper analyzes characteristic points in flat channels (e.g., rectangular channel) and concave boundary channels (e.g., semi-circular channel). In the rectangular channel, the division line divides the section into three regions. In each region, the analysis is conducted in the direction perpendicular to the bottom or side wall of the channel. In the semi-circular channel, the analysis is conducted along the normal direction. Based on the log-law, the theoretical expressions for calculating the location of the average velocity characteristic points in flat and concave boundary channels are derived through the formula transformation.

(3) The velocity data in different experimental sites are used to verify the validity of the CPL formulas applied to flat and concave boundary channels. Moreover, the discharge calculation formulas of channels are given through discussion with CPL.

6. Conclusions

The determination of CPL has provided a theoretical basis for simplifying and accurately measuring channel flow. It is precisely because of the realization of CPL that the average cross-section discharge can be calculated quickly so that the accuracy and stability of flow measurement in irrigation channels can be further improved.

The results of this paper can only serve for smooth open channels. However, if the roughness is non-uniform, the location of the division lines in the channel cross-section will be different [25]. It is also difficult to ensure that the velocity distribution in each sub-region is the same. With the comprehensive ecological treatment of channels becoming a hot topic, the transition from smooth channels to vegetation covered channels becomes relevant. So, it is of necessity to conduct experiments in open channels with roughness for future research.

Author Contributions: Conceptualization, Y.H., J.C., and T.L.; Methodology, Z.M., J.C., and T.L.; Software, J.W., Z.M., J.C., and T.L.; Validation, Y.H., Z.M., J.W., and T.L.; Writing—Original Draft Preparation, T.L. and Y.H.; Project Administration, Y.H.; Funding Acquisition, Y.H. and J.C. All authors have read and agreed to the published version of the manuscript.

Funding: The authors express gratitude for the financial support. This research was funded by the National Natural Science Foundation of China, grant No. 51979275, by the National Key Research and Development Program of China under grant Nos. 2017YFC0403203 and 2017YFD0701003, by the Jilin Province Key Research and Development Plan Project under grant No. 20180201036SF, by the Open Fund of Synergistic Innovation Center of Jiangsu Modern Agricultural Equipment and Technology, Jiangsu University, under grant No. 4091600015, by the Open Research Fund of State Key Laboratory of Information Engineering in Surveying, Mapping and Remote Sensing, Wuhan University, under grant No. 19R06, and by the Chinese Universities Scientific Fund under grant No. 10710301.

Acknowledgments: The authors are very grateful for the funding support.

Conflicts of Interest: The authors declare no conflict of interest.

References

1. Mancosu, N.; Snyder, R.L.; Kyriakakis, G.; Spano, D. Water Scarcity and Future Challenges for Food Production. *Water* **2015**, *7*, 975–992. [CrossRef]
2. Salehi, S.; Azimi, A.H. Discharge Characteristics of Weir-Orifice and Weir-Gate Structures. *J. Irrig. Drain. Eng.* **2019**, *145*. [CrossRef]
3. Samani, Z. Three simple flumes for flow measurement in open channels. *J. Irrig. Drain. Eng.* **2017**, *143*. [CrossRef]
4. Samani, Z.; Magallanez, H. Simple flume for flow measurement in open channel. *J. Irrig. Drain. Eng.* **2000**, *126*, 127–129. [CrossRef]
5. Keulegan, G.H. Laws of turbulent flow in open channels. *J. Res. Natl. Bur. Stand.* **1938**, *21*, 707–741. [CrossRef]
6. Zanoun, E.-S.; Durst, F.; Bayoumy, O.; Al-Salaymehc, A. Wall skin friction and mean velocity profiles of fully developed turbulent pipe flows. *Exp. Therm. Fluid. Sci.* **2007**, *32*, 249–261. [CrossRef]
7. Nikuradse, J. *Laws of Flow in Rough Pipes (1950 translation)*; National Advisory Committee for Aeronautics: Washington, DC, USA, 1933; NACA-TM-1292. Available online: https://ntrs.nasa.gov/search.jsp?R=19930093938 (accessed on 15 January 2020).
8. Cardoso, A.H.; Graf, W.H.; Gust, G. Uniform-Flow in a Smooth Open Channel. *J. Hydraul. Res.* **1989**, *27*, 606–616. [CrossRef]
9. Steffler, P.M.; Rajaratnam, N.; Peterson, A.W. LDA Measurements in Open Channel. *J. Hydraul. Eng.-ASCE* **1985**, *111*. [CrossRef]
10. Nezu, I.; Rodi, W. Open-Channel Flow Measurements with a Laser Doppler Anemometer. *J. Hydraul. Eng.* **1986**, *112*, 335–355. [CrossRef]
11. Chiu, C.L. Entropy and 2-D velocity distribution in open channels. *J. Hydraul. Eng.* **1988**, *114*, 738–756. [CrossRef]
12. Chiu, C.L. Application of entropy concept in open-channel flow study. *J. Hydraul. Eng.* **1991**, *117*, 615–628. [CrossRef]
13. Chiu, C.L.; Murray, D.W. Variation of velocity distribution along non-uniform open-channel flow. *J. Hydraul. Eng.* **1992**, *118*, 989–1001.
14. Maghrebi, M.F. Application of the single point measurement in discharge estimation. *Adv. Water Resour.* **2006**, *29*, 1504–1514. [CrossRef]
15. Bretheim, J.U.; Meneveau, C.; Gayme, D.F. Standard logarithmic mean velocity distribution in a band-limited restricted nonlinear model of turbulent flow in a half-channel. *Phys. Fluids* **2015**, *27*. [CrossRef]
16. Hong, J.H.; Guo, W.D.; Wang, H.W.; Yeh, P.H. Estimating discharge in gravel-bed river using non-contact ground-penetrating and surface-velocity radars. *River Res. Appl.* **2017**, *33*, 1177–1190. [CrossRef]
17. Moramarco, T.; Barbetta, S.; Tarpanelli, A. From Surface Flow Velocity Measurements to Discharge Assessment by the Entropy Theory. *Water* **2017**, *9*, 120. [CrossRef]
18. Johnson, E.D.; Cowen, E.A. Estimating bed shear stress from remotely measured surface turbulent dissipation fields in open channel flows. *Water Resour. Res.* **2017**, *53*, 1982–1996. [CrossRef]

19. Johnson, E.D.; Cowen, E.A. Remote determination of the velocity index and mean streamwise velocity profiles. *Water Resour. Res.* **2017**, *53*, 7521–7535. [CrossRef]
20. Khuntia, J.R.; Kamalini, D.; Khatua, K.K. Prediction of depth-averaged velocity in an open channel flow. *Appl. Water Sci.* **2018**, *8*, 172. [CrossRef]
21. Han, Y.; Yang, S.Q.; Dharmasiri, N.; Sivakumar, M. Experimental study of smooth channel flow division based on velocity distribution. *J. Hydraul. Eng.* **2015**, *141*. [CrossRef]
22. Yang, S.Q.; Han, Y.; Lin, P.Z.; Jiang, C.B.; Walker, R. Experimental study on the validity of flow region division. *J. Hydro-Environ. Res.* **2014**, *8*, 421–427. [CrossRef]
23. Melling, A.; Whitelaw, J.H. Turbulent flow in a rectangular duct. *J. Fluid Mech.* **1976**, *78*, 289–315. [CrossRef]
24. Tracy, H.J. Turbulent flow in a three-dimensional channel. *J. Hydraul. Div.* **1965**, *91*, 9–35.
25. Einstein, H.A. Formulas for the transportation of bed load. *Trans. ASCE* **1942**, *107*, 133–169.
26. Chien, N.; Wang, Z. *Mechanics of Sediment Transport*; Science Press: Beijing, China, 1986. (In Chinese)
27. Yang, S.Q.; Lim, S.Y. Mechanism of energy transportation and turbulent flow in a 3D channel. *J. Hydraul. Eng.-ASCE* **1997**, *123*, 684–692. [CrossRef]
28. Luo, J. Study on the Hydraulic Characteristics of Pressure-Free Uniform Flow in Circular Section Pipeline. Master's Thesis, Tsinghua University, Beijing, China, 2016. (In Chinese).
29. Knight, D.W.; Sterling, M. Boundary shear in circular pipes running partially full. *J. Hydraul. Eng.-ASCE* **2000**, *126*, 263–275. [CrossRef]
30. Chiu, C.L.; Chiou, J.D. Structure of 3-D flow in rectangular open channels. *J. Hydraul. Eng.-ASCE* **1986**, *112*, 1050–1068. [CrossRef]
31. Stearns, F.P. A reason why the maximum velocity of water flowing in open channels is below the surface. *Trans. Am. Soc. Civ. Eng.* **1883**, *7*, 331–338.
32. Murphy, C. Accuracy of stream measurements. *Water Supply Irrig.* **1904**, *95*, 111–112.
33. Tominaga, A.; Nezu, I. Turbulent structure in compound open-channel flows. *J. Hydraul. Eng.-ASCE* **1991**, *117*, 21–41. [CrossRef]
34. Mirauda, D.; Pannone, M.; De Vincenzo, A. An Entropic Model for the Assessment of Streamwise Velocity Dip in Wide Open Channels. *Entropy.* **2018**, *20*, 69. [CrossRef]
35. Coles, D. The Law of the Wake in the Turbulent Boundary Layer. *J. Fluid Mech.* **1956**, *1*, 191–226. [CrossRef]
36. Absi, R. An ordinary differential equation for velocity distribution and dip-phenomenon in open channel flows. *J. Hydraul. Res.* **2011**, *49*, 82–89. [CrossRef]

Review

Potential Fields in Fluid Mechanics: A Review of Two Classical Approaches and Related Recent Advances

Markus Scholle [1,*,†], Florian Marner [2,†] and Philip H. Gaskell [3,†]

1 Department of Mechatronics and Robotics, Heilbronn University, D-74081 Heilbronn, Germany
2 Inigence GmbH, D-74626 Bretzfeld, Germany; florian_marner@gmx.net
3 Department of Engineering, Durham University, Durham DH1 3LE, UK; p.h.gaskell@durham.ac.uk
* Correspondence: markus.scholle@hs-heilbronn.de
† These authors contributed equally to this work.

Received: 24 March 2020; Accepted: 24 April 2020; Published: 27 April 2020

Abstract: The use of potential fields in fluid dynamics is retraced, ranging from classical potential theory to recent developments in this evergreen research field. The focus is centred on two major approaches and their advancements: (i) the Clebsch transformation and (ii) the classical complex variable method utilising Airy's stress function, which can be generalised to a first integral methodology based on the introduction of a tensor potential and parallels drawn with Maxwell's theory. Basic questions relating to the existence and gauge freedoms of the potential fields and the satisfaction of the boundary conditions required for closure are addressed; with respect to (i), the properties of self-adjointness and Galilean invariance are of particular interest. The application and use of both approaches is explored through the solution of four purposely selected problems; three of which are tractable analytically, the fourth requiring a numerical solution. In all cases, the results obtained are found to be in excellent agreement with corresponding solutions available in the open literature.

Keywords: potential fields; Clebsch variables; Airy's stress function; Goursat functions; Galilean invariance; variational principles; boundary conditions; film flows

PACS: 47.10.-g

MSC: 76M99

---

1. Introduction

In various branches of physics, potentials continue to be used as additional auxiliary fields for the advantageous reformulation of one or more governing equations. Classical electrodynamics serves as a prime illustrative example of their use [1], enabling Maxwell equations in their commonly expressed form, comprised of two scalar and two vector equations for the two observables—electric field $\vec{E}$ and magnetic flux density $\vec{B}$—to be formulated differently in terms of the derivatives of a scalar potential φ and a vector potential $\vec{A}$, such that $\vec{E} = -\nabla\varphi - \partial_t\vec{A}$ and $\vec{B} = \nabla \times \vec{A}$, respectively; in which case, two of the four Maxwell equations are fulfilled identically while the other two form a self-adjoint pair—i.e., can be obtained from a variational principle. Moreover, followed by proper gauging of the potential fields, a fully decoupled form of Maxwell's equations in terms of two inhomogeneous d'Alembert equations is obtained. From a mathematical viewpoint, amendments to the original equations comprise:

- A reduction in the number of equations and unknowns;
- Self-adjointness of the equations;
- A full decoupling of the equations;
- Transformation of the equations to a known mathematical type.

Despite the above benefits, a historical debate reaching back to Heaviside [2,3] has surrounded the Maxwell potentials and their rationale, which were not free of polemics [4–6]. A major objection against the use of potential fields has surrounded their mathematical non-uniqueness, implying that they are not observable by physical measurements. In the case of the Maxwell potentials, the transformation

$$\vec{A} \rightarrow \vec{A}' = \vec{A} + \nabla\chi \tag{1}$$

$$\varphi \rightarrow \varphi' = \varphi - \partial_t\chi \tag{2}$$

in terms of an arbitrary scalar field χ defines an alternative set of potentials related identically to the observable fields $\vec{E}$ and $\vec{B}$. The non-uniqueness of the original Maxwell potentials does not reflect a weak point in the theory; on the contrary, it defines a symmetry that has proven formative in the usage of potential fields in modern theoretical physics as well as ground-breaking with respect to the subsequent development of gauge theory [7]. Each of which confirm Maxwell's theory to be a significant reference point for other field theories, an obvious example being the potential representation of the Weyl tensor in general relativity [8,9].

Subsequently, based on the publications of Ehrenberg and Siday [10], Aharonov and Bohm [11], the so-called *Aharonov–Bohm solenoid effect*—which takes place when the wave function of a charged particle passes around a long solenoid and experiences a phase shift as a result of the enclosed magnetic field, despite the magnetic field being negligible outside the solenoid—gave rise to a second debate concerning the physical meaning of the vector Potential $\vec{A}$. This effect is frequently misunderstood: it does not allow for a point-wise "measurement" of $\vec{A}$ since only the integral magnetic flux can be determined from the phase shift, while the gauge transformation in Equation (1) is still a symmetry of the Schrödinger–Maxwell theory predicting the effect properly; thus any gradient field $\nabla\chi$ can be added to the vector potential $\vec{A}$ leading to the same experimental result. Nevertheless, the question still remains an open one as to what physical role the vector potential plays and controversy surrounding it persists; see, for example, [12–18].

In summary, the use of potentials is motivated from both a mathematical and physical viewpoint: mathematically the original equations can be manipulated in beneficial ways, while physically new insight concerning the structure of the theory is provided via the reformulated equation set. Both of these aspects are considered subsequently.

The particular aim henceforth is to demonstrate the application of potential methods to fluid mechanics. As is well known in the case of an incompressible inviscid irrotational fluid flow, by expressing the velocity field $\vec{v}$ in terms of a scalar potential φ:

$$\vec{v} = \nabla\varphi \tag{3}$$

the associated equations of motion are reduced to Bernoulli's equation, resulting as a *first integral* of Euler's equations together with a Laplace equation

$$\nabla^2\varphi = 0 \tag{4}$$

for the potential via the continuity equation. Since Bernoulli's equation is essentially the conditional equation for the pressure field, only Equation (4) remains to be solved, elevating potential flow theory to the status of an essential topic in standard fluid dynamics text books [19–22]. Despite the obvious advantage of making various flow problems more tractable, the approach is restricted to inviscid and irrotational barotropic flows. In order to extend the methodology to encompass a wider range of flow problems, generalisations of and alternatives to Equation (3) have emerged. Of these, the following two major strands only are explored in the present work:

1. The so-called *Clebsch transformation* [19,20,23] and related methodologies enabling, in the first instance, Euler's equation to be reformulated as a generalised Bernoulli equation complemented with two transport equations for the Clebsch potentials. The approach was subsequently generalised to encompass baroclinic flow by Seliger and Witham [24], but still with the restriction that the flow is inviscid and heat conduction absent. Note that the Clebsch transformation has been applied to physical problems beyond fluid mechanics including Maxwell theory in classical electrodynamics [25], the field of Magnetohydrodynamics [26], relativistic dynamical systems [27] and even in relation to quantum theory within the context of (a) the quantisation of vortex tubes Madelung [28], Schoenberg [29], (b) generalised membranes [30] and (c) relativistic quantum vorticity [31].
2. The *complex variable method*, developed in the first half of the 20th century and originally related to problems in plane linear elasticity [32,33]. The method was subsequently adopted by the fluid mechanics community: in the case of 2D Stokes flow (Re $\to$ 0) it has led to solutions based on a complex-valued Goursat representation of the stream function in terms of two analytic functions, which has been generalised incrementally, starting with Legendre [34] and followed by Coleman [35], Ranger [36], and Scholle et al. [37], Marner et al. [38], resulting finally in an exact complex-valued first integral of the 2D unsteady NS equations, based on the introduction of an auxiliary potential field. A further generalisation to 3D viscous flow has been achieved only recently using a tensor potential in place of the complex potential field employed in two-dimensions [39].

Historically, both approaches have been subject to limitations on their usage: the Clebsch transformation originated for the case of inviscid flow (Re $\to \infty$) and the complex variable method for that of 2D Stokes flow (Re $\to$ 0). Recent advancements have lifted these restrictions. In this review, the origin and evolution of both methods is retraced and potential future developments highlighted. Section 2 considers the Clebsch transformation, starting from its origins in inviscid barotropic flow theory, Section 2.1. After commenting on the global existence of the Clebsch variables, Section 2.2, it is demonstrated via a rigorous symmetry analysis, Section 2.2, that the Clebsch representation of the velocity is a natural outcome of Galilean invariance via Noether's theorem. An extended Clebsch transformation for viscous flow is then presented in Section 2.3 and followed, Section 2.4, by applying it to the problem of viscous stagnation flow. The complex variable method and its progression to a tensor potential approach is outlined in Section 3, beginning in Section 3.1 with the use of Airy's stress function with respect to steady-state equilibrium conditions for an arbitrary continuum in general and for Stokes's flow in particular. In Section 3.2, the approach is generalised to encompass the full unsteady 2D-NS equations, utilising a complex potential. A serendipitous benefit is that of enabling the integration of the dynamic boundary condition along a free surface, or interface, and its reduction to a standard form, as shown in Section 3.3, as subsequently utilised in Section 3.4 for the numerical solution of a free surface film flow problem. Latterly, a particularly noteworthy advance, Section 3.5, has been that of overcoming the 2D restriction. This has been achieved by rearranging the complex equations in a tensor form leading to a potential-based first integral of the full 3D-NS equations with a seamless extension to an analogous form of the 4D relativistic energy momentum equations. Concluding remarks are provided in Section 4.

Finally, it would be remiss and incomplete not to point out that in the field of fluid mechanics other approaches to the use of potentials for solving the equations of motion exist that have not been considered in the present text. In this sense and as instructive examples, the reader is referred to the work of Papkovich and Neuber [40], Lee et al. [41], Greengard and Jiang [42].

2. Clebsch Transformation Approach

2.1. The Clebsch Transformation for Inviscid Flows

For inviscid flow, Clebsch [19,20,23] proposed a non-standard potential representation for the velocity field, in terms of the so-called Clebsch variables φ, α, β, of the form:

$$\vec{u} = \nabla\varphi + \alpha\nabla\beta\,. \tag{5}$$

From a mathematical viewpoint, the potential representation in Equation (5) is a decomposition of the velocity field into a curl-free part $\nabla\varphi$ and a helicity-free part $\alpha\nabla\beta$. Schoenberg [29] has shown that the above decomposition is not unique; by applying the gauge transformation:

$$\begin{array}{lcll} \varphi & \longrightarrow & \varphi' & = \varphi + f(\alpha,\beta,t) \\ \alpha & \longrightarrow & \alpha' & = g(\alpha,\beta,t) \\ \beta & \longrightarrow & \beta' & = h(\alpha,\beta,t) \end{array} \tag{6}$$

an equivalent set of Clebsch variables $\varphi', \alpha', \beta'$ is given if and only if the functions f, g, h fulfil the following two PDEs:

$$\frac{\partial f}{\partial \beta} + g\frac{\partial h}{\partial \beta} = \alpha\,, \tag{7}$$

$$\frac{\partial f}{\partial \alpha} + g\frac{\partial h}{\partial \alpha} = 0\,. \tag{8}$$

By applying Equation (5) to Euler's equations for inviscid flows, one obtains:

$$\vec{0} = \frac{\mathrm{D}\vec{u}}{\mathrm{D}t} + \nabla[P+U] = \nabla\left[\frac{\partial\varphi}{\partial t} + \alpha\frac{\partial\beta}{\partial t} + \frac{\vec{u}^2}{2} + P(\varrho) + U\right] + \frac{\mathrm{D}\alpha}{\mathrm{D}t}\nabla\beta - \frac{\mathrm{D}\beta}{\mathrm{D}t}\nabla\alpha\,, \tag{9}$$

with the pressure function $P(\varrho) = \int \varrho^{-1}\mathrm{d}p$ and U the specific potential energy of the external force. The operator $\mathrm{D}/\mathrm{D}t = \partial/\partial t + \vec{u}\cdot\nabla$ is the material time derivative. Being basically of the form:

$$\nabla[\cdots] + [\cdots]\nabla\alpha + [\cdots]\nabla\beta = \vec{0}, \tag{10}$$

this vector equation can be decomposed according to:

$$\frac{\partial\varphi}{\partial t} + \alpha\frac{\partial\beta}{\partial t} + \frac{\vec{u}^2}{2} + P + U = F(\alpha,\beta,t), \tag{11}$$

$$\frac{\mathrm{D}\alpha}{\mathrm{D}t} = -\frac{\partial F}{\partial\beta}, \tag{12}$$

$$\frac{\mathrm{D}\beta}{\mathrm{D}t} = \frac{\partial F}{\partial\alpha}, \tag{13}$$

containing an unknown function $F(\alpha,\beta,t)$; which, making use of the gauge transformation of Equation (6), leads to $F \to 0$. The above three scalar field equations represent a first integral of Euler's equations and are self-adjoint. However, their most intriguing feature is that the vorticity:

$$\vec{\omega} = \frac{1}{2}\nabla\times\vec{u} = \frac{1}{2}\nabla\alpha\times\nabla\beta, \tag{14}$$

is given by the two scalar fields α, β, only. Hence, the vortex dynamics is reduced to and conveniently captured by the two transport Equations (12) and (13). This beneficial feature has been exploited by, for example, Prakash et al. [43], who utilised the Clebsch transformation as a generalisation of classical potential theory for the simulation of bubble dynamics.

Another beneficial feature of the transformed field equations is their *self-adjointness*: they result from a variational principle:

$$\delta \int_{t_1}^{t_2} \iiint_V \ell\left(\varrho, \alpha, \partial\varphi, \partial\beta, \vec{x}, t\right) \mathrm{d}V \mathrm{d}t = 0, \tag{15}$$

with respect to free and independent variation of the four fields $\varrho, \varphi, \alpha, \beta$ and their first order spatial and temporal derivatives for fixed values at initial and final time, $t_{1,2}$, with the Lagrangian given as [25]:

$$\ell\left(\varrho, \alpha, \partial\varphi, \partial\beta, \vec{x}, t\right) = -\varrho\left[\frac{\partial\varphi}{\partial t} + \alpha\frac{\partial\beta}{\partial t} + \frac{1}{2}\left(\nabla\varphi + \alpha\nabla\beta\right)^2 + e\left(\varrho\right) + U\left(\vec{x}, t\right)\right], \tag{16}$$

for the specific elastic energy $e(\varrho)$ fulfilling $e(\varrho) + \varrho e'(\varrho) = P(\varrho)$. For generalisations of the above variational principle toward thermal degrees of freedom, reference is made to [24,44–46].

2.2. A Note on the Global Existence of the Clebsch Variables

When introducing potentials as auxiliary fields, their existence has to be clarified first. Referring to the classical example $\vec{u} = \nabla\varphi$ from potential theory corresponding to the special case of the Clebsch representation in Equation (5) with $\alpha = \beta = 0$, it is obvious that it applies only to vortex-free fields, i.e., $\nabla \times \vec{u} = \vec{0}$. However, according to Equation (14), velocity fields with non-vanishing vorticity (Equation ((14)) can be considered, but not arbitrary ones, as demonstrated in the following, Moreau [47], Moffatt [48] having identified the *helicity*

$$H = \iiint_V 2\vec{u} \cdot \vec{\omega} \mathrm{d}V \tag{17}$$

as a decisive quantity in the case of inviscid fluid flow. Under the assumption that the potential φ is continuously differentiable and single-valued then by expressing, via Equations (5) and (14), the velocity and the vorticity in terms of the Clebsch variables and their gradients, the helicity density can be re-written as:

$$\begin{aligned} 2\vec{u} \cdot \vec{\omega} &= \left[\nabla\varphi + \alpha\nabla\beta\right] \cdot \left(\nabla\alpha \times \nabla\beta\right) = \nabla\varphi \cdot \left(\nabla\alpha \times \nabla\beta\right), \\ &= \nabla\varphi \cdot \vec{\omega} = 2\nabla \cdot \left(\varphi\vec{\omega}\right) - 2\varphi \underbrace{\nabla \cdot \vec{\omega}}_{\vec{0}} \end{aligned}$$

implying, utilising Gauss's theorem with ∂V denoting the surface of the flow domain and $\vec{n}$ the normal vector, the global helicity to be:

$$H = 2\iiint_V \nabla \cdot \left(\varphi\vec{\omega}\right) \mathrm{d}V = \oiint_{\partial V} \varphi\vec{\omega} \cdot \vec{n} \mathrm{d}A\,. \tag{18}$$

The issue raised by the above formula is the occurrence of the potential φ as a non-observable in the sense that it can, according to Equation (6), be replaced by a re-gauged potential $\varphi' = \varphi + f(\alpha, \beta)$, leading to a corresponding helicity given by:

$$H' = H + 2\oiint_{\partial V} f\left(\alpha, \beta\right) \vec{\omega} \cdot \vec{n} \mathrm{d}A\,. \tag{19}$$

Since, due to its definition (Equation ((17)), the helicity is an observable and therefore a gauge-invariant quantity, i.e., $H' = H$, the integral on the right hand of Equation (19) has to vanish for any choice of the function $f(\alpha, \beta)$, implying that $\vec{\omega} \cdot \vec{n} = 0$ along the entire boundary ∂V of the flow domain. Finally, the latter implies also that the helicity vanishes. As a consequence, the classical Clebsch transformation with continuously differentiable and single-valued potentials only applies to flows the total helicity of which is zero.

There are two different approaches to overcoming this restriction: the first utilises a multi-valued potential φ, as demonstrated by Yahalom [49], Yahalom and Lynden-Bell [50]; the second is based on the use of multiple pairs of variables, such as $\vec{u} = \nabla\varphi + \alpha_1\nabla\beta_1 + \alpha_2\nabla\beta_2 + \cdots$, in the sense that the number of pairs has to be chosen adequately depending on the topological features of the respective individual flow problem—such flows with closed vortex lines that form linked rings or ones with isolated points of zero vorticity are discussed, for example, by Balkovsky [51], Yoshida [52]. More recent work on this topic is presented by Ohkitani and Constantin [53], Cartes et al. [54], Ohkitani [55]. However, independently of the question of how many pairs of Clebsch variables are useful for representing the topology of a flow, there is a minimum number of two pairs for which global existence can be granted, as demonstrated in Appendix A.

Subsequently, for convenience, attention is paid to the classical form in Equation (5) only on the understanding that an extension via more pairs of variables is possible if required.

Derivation of a Clebsch-Like Form by Galilean Invariance and Self-Adjointness

The question of global existence discussed briefly above, Section 2.2, has become a research topic to which many research groups have contributed over several decades, while the use of the Clebsch variables remained restricted to the realm of inviscid flows. In contrast, the focus here is to provide a rationale for the use of Clebsch variables for arbitrary continuous physical systems, the dynamics of which are assumed to be given by a variational principle, following an in-depth analysis of the underlying Galilean group and its consequences for the resulting balances of mass and momentum.

If a system is *physically closed*, i.e., isolated from the surrounding, its equations of motion are invariant with respect to the following four universal symmetry transformations of the Galilean group, corresponding to homogeneity of time and space, isotropy of space and equality of all inertial frames:

time translations:

$$t \rightarrow t' = t + \tau \tag{20}$$

space translations:

$$\vec{x} \rightarrow \vec{x}' = \vec{x} + \vec{s} \tag{21}$$

rigid rotations:

$$\vec{x} \rightarrow \vec{x}' = \underline{R}\vec{x} \tag{22}$$

Galilei boosts:

$$\vec{x} \rightarrow \vec{x}' = \vec{x} - \vec{u}_0 t \tag{23}$$

Here, the scalar τ, the two vectors $\vec{s}$ and $\vec{u}_0$ and the unitary matrix $\underline{R}$ fulfilling $\underline{R}^T\underline{R} = \underline{1}$ and $\det \underline{R} = 1$ are constants. Via Formulae (20)–(23) the four symmetries are obviously well-defined for discrete systems; for instance, systems of point masses in Newtonian mechanics. For continuous systems, the situation is essentially different: in field theories, the Formulae (20)–(23) have to be supplemented by the respective transformation formulae for the fields in order to define the transformations completely. For demonstration purposes consider the Lagrangian (16) for inviscid barotropic flows in the absence of external forces, $U = 0$:

$$\ell(\varrho, \alpha, \dot{\varphi}, \nabla\varphi, \dot{\beta}, \nabla\beta) = -\varrho\left[\dot{\varphi} + \alpha\dot{\beta} + \frac{1}{2}(\nabla\varphi + \alpha\nabla\beta)^2 + e(\varrho)\right], \tag{24}$$

where the dot indicates the partial time derivative $\partial/\partial t$. Obviously, the Lagrangian (24) is invariant with respect to space and time translations in Equations (20), (21) and rigid rotations (22) if the four fields, $\varrho, \varphi, \alpha, \beta$ are assumed to be likewise invariant. In contrast, invariance with respect to Galilei boosts requires the first Clebsch variable φ to be transformed according to [26]:

$$\varphi \to \varphi' = \varphi - \vec{u}_0 \cdot \vec{x} + \frac{\vec{u}_0^2}{2} t, \tag{25}$$

implying, according to Equation (5), that consequently $\vec{u} \to \vec{u}' = \vec{u} - \vec{u}_0$ for the velocity field and compensating the non-invariance $\partial/\partial t \to \partial/\partial t + \vec{u}_0 \cdot \nabla$ of the partial time derivatives occurring in the Lagrangian. The fact that the Galilei boost becomes manifest as a combination of the geometric transformation in Equation (23) with the gauge transformation in Equation (25) is an unfavourable feature since, for each physical system depending on an arbitrary set of fields, the transformation formulae for the fields involved have to be defined individually. However, there is a way to eliminate this disadvantage: by means of the substitution:

$$\varphi = \Phi + \zeta, \tag{26}$$

with the *generating field*:

$$\zeta := \frac{\vec{x}^2}{2t}. \tag{27}$$

The Lagrangian (24) can be re-written in the alternative form:

$$\tilde{\ell}\left(\varrho, \alpha, \overset{\circ}{\Phi}, \nabla\Phi, \overset{\circ}{\beta}, \nabla\beta\right) = -\varrho\left[\overset{\circ}{\Phi} + \alpha\,\overset{\circ}{\beta} + \frac{1}{2}\left(\nabla\varphi + \alpha\nabla\beta\right)^2 + e\left(\varrho\right)\right], \tag{28}$$

where the ring symbol $\circ$ indicates the *dual time derivative* [56]:

$$\overset{\circ}{\Phi} = \left\{\frac{\partial}{\partial t} + \nabla\zeta \cdot \nabla\right\}\Phi, \tag{29}$$

which in contrast to the conventional time derivative is invariant with respect to Galilei boosts. As a consequence, the Lagrangian in its alternative form in Equation (28), subsequently called the *dual representation*, proves to be invariant with respect to Galilei boosts if all fields including Φ are assumed to be likewise invariant. Thus, the essence of the dual representation in Equation(28) is that Galilei boosts become manifest as pure geometrical transformations without the need to combine them with a re-gauging of potentials. Vice versa, for the dual Lagrangian (28), time and space translations become manifest as mixed transformations consisting of a geometric part in combination with a re-gauging of the potential Φ [56]. Subsequently, a Lagrangian is termed *strictly invariant* if it is invariant with respect to a purely geometric transformation without the need of re-gauging one of the potential fields. In this sense, Equation (24) is strictly invariant with respect to space and time translations while Equation (28) is strictly invariant with respect to Galilei boosts but not vice versa.

This poses the question of whether the coexistence of two different representations of a given Lagrangian, one being strictly invariant with respect to space and time translations and the other one being strictly invariant with respect to Galilei boosts, is an individual feature of the theory of inviscid barotropic flows or if other examples exist. Scholle [56] undertook a rigorous analysis to investigate this question, arriving at the conclusion that this coexistence is a *universal property of every continuum* being ruled by self-adjoint equations of motions invariant with respect to the full Galilean group: if the state of a continuous system is defined by N independent fields $\psi_i, i = 1 \cdots, N$ and its evolution determined by a variational principle:

$$\delta \int_{t_1}^{t_2} \iiint_V \ell\left(\psi_i, \dot{\psi}_i, \nabla\psi_i\right) \mathrm{d}V\mathrm{d}t = 0\,, \tag{30}$$

based on free and independent variation of the fields ψ_i and their first order spatial and temporal derivatives for fixed values at the initial and final time, $t_{1,2}$, then a variable transformation

$$\psi_i = K_i\left(\Psi_j, \zeta, \nabla\zeta\right)\,, \tag{31}$$

exists with the generating field ζ defined via Equation (27), fulfilling the identity:

$$\ell\left(K_i\left(\Psi_j,\zeta,\nabla\zeta\right), \dot{K}_i\left(\Psi_j,\zeta,\nabla\zeta\right), \nabla K_i\left(\Psi_j,\zeta,\nabla\zeta\right)\right) = \ell\left(\Psi_i, \overset{\circ}{\Psi}_i, \nabla\Psi_i + \frac{1}{t}\vec{K}_i\left(\Psi_j\right)\right)\,, \tag{32}$$

$$\vec{K}_i\left(\Psi_j\right) := \lim_{\zeta,\nabla\zeta\to 0} \frac{\partial K_i}{\partial(\nabla\zeta)}\,, \tag{33}$$

giving rise to the dual representation of the Lagrangian on the right hand of Equation (32) depending on the dual time derivatives $\overset{\circ}{\Psi}_i = \dot{\Psi}_i + \nabla\zeta\cdot\nabla\Psi_i$ of the transformed fields.

Since the conventional representation $\ell\left(\psi_i, \dot{\psi}_i, \nabla\psi_i\right)$ is obviously strictly invariant with respect to space and time translations while the dual representation $\ell\left(\Psi_i, \overset{\circ}{\Psi}_i, \nabla\Psi_i + \vec{K}_i\left(\Psi_j\right)/t\right)$ is strictly invariant with respect to Galilei boosts, simultaneous invariance with respect to translations and Galilei boosts is granted by Equation (32), which consequentially can be understood as a collective symmetry criterion for the Galilean group and has to be fulfilled by any Lagrangian related to a physically closed continuous system. One implication is that the gauge transformation:

$$\psi_i \to \psi_i{}' = K_i\left(\psi_j, \varepsilon, 0\right) \tag{34}$$

is likewise defined as being a symmetry transformation of the Lagrangian. This *induced gauge transformation* is remarkable since it is an indirect consequence of the Galilean invariance.

The above-mentioned general properties of Lagrangians in continuum theory entail additional general implications for the physical balances resulting from the variational Principle (30) by utilising Noether's theorem [57–59], the essence of which is that each Lie symmetry of the Lagrangian gives rise to a physical balance equation and to a canonical definition of the density and flux density involved. In the present context, the balances for mass and momentum,

$$\frac{\partial\varrho}{\partial t} + \nabla\cdot(\varrho\vec{u}) = 0\,, \tag{35}$$

$$\frac{\partial\vec{p}}{\partial t} + \nabla\cdot\underline{\Pi} = \vec{0}\,, \tag{36}$$

are related (via Noether's theorem) to the induced gauge Transformation (34) and (from the space translation in Equation (21)) with the mass density ϱ, the mass flux density $\vec{j} = \varrho\vec{u}$, the momentum density $\vec{p}$ and the momentum flux density $\underline{\Pi}$ which are given by [56]:

$$\varrho = -\frac{\partial \ell}{\partial \dot{\psi}_i} K_{0i}\left(\psi_j\right) \; , \tag{37}$$

$$\vec{j} = \varrho\vec{u} = -\frac{\partial \ell}{\partial \nabla \psi_i} K_{0i}\left(\psi_j\right) \; , \tag{38}$$

$$\vec{p} = -\frac{\partial \ell}{\partial \dot{\psi}_i} \nabla \psi_i \, , \tag{39}$$

$$\underline{\Pi} = \ell \underline{1} - \frac{\partial \ell}{\partial \nabla \psi_i} \otimes \nabla \psi_i \, , \tag{40}$$

with the infinitesimal Generator $K_{0i}\left(\psi_j\right)$ defined as:

$$K_{0i}\left(\Psi_j\right) := \lim_{\zeta, \nabla\zeta \to 0} \frac{\partial K_i}{\partial \zeta} \, . \tag{41}$$

In Equations (37)–(40) and subsequently, the following *Einstein notation* is utilised: an index variable appearing twice in a single term implies summation of that term over all values of the index.

Note that in contrast to classical continuum mechanics, the mass flux density and the momentum density need not to be equal. The difference between both, $\vec{p}^* := \vec{p} - \vec{j} = \vec{p} - \varrho\vec{u}$, is called the *quasi-momentum density* and can be interpreted as contributions to the momentum due to non-material degrees of freedom, for example, electromagnetic fields, thermal fields and also due to phenomena beyond the scope of continuum theories on a molecular scale, for example, Brownian motion. After analysing the Noether balance resulting from Galilei-boosts, a constitutive relation between momentum and mass flux can be identified [56] implying the identity:

$$\vec{p}^* = -\frac{\partial}{\partial t}\left[\frac{\partial \ell}{\partial \dot{\psi}_i}\vec{K}_i\left(\psi_j\right)\right] - \nabla \cdot \left[\frac{\partial \ell}{\partial \nabla \psi_i} \otimes \vec{K}_i\left(\psi_j\right)\right] \, . \tag{42}$$

for the quasi-momentum density.

In classical continuum mechanics, mass flux density and momentum density are assumed to be equal, implying $\vec{p}^* = \vec{0}$, which according to Equations (42) and (33) requires the variable Transformation (31) to be independent of $\nabla\zeta$, leading to a drastic mathematical simplification of Criterion (32) and allowing derivation of the following universal scheme for Lagrangians [56]: without loss of generality the conventional representation of the Lagrangian can be written in terms of N fields, $(\psi_i) = (\varphi, \vartheta_1, \cdots, \vartheta_{N-1})$ as:

$$\ell\left(\dot{\varphi}, \nabla\varphi, \vartheta^j, \dot{\vartheta}^j, \nabla\vartheta^j\right) = L\left(\omega, \vartheta_j, \overset{\odot}{\vartheta}_j, \nabla\vartheta_j\right), \tag{43}$$

with the abbreviations:

$$\omega := \dot{\varphi} + \frac{1}{2}\left(\nabla\varphi\right)^2 \, , \tag{44}$$

$$\overset{\odot}{\vartheta}_j := \dot{\vartheta}_j + \nabla\varphi \cdot \nabla\vartheta_j \, , \qquad j = 1, \cdots, N-1 \, , \tag{45}$$

fulfilling Criterion (32) for the following form of the variable Transformation (31):

$$\varphi = \Phi + \zeta \, , \tag{46}$$

$$\vartheta_j = \Theta_j \, , \qquad j = 1, \cdots, N-1 \, . \tag{47}$$

Consequently, the Lagrangian (24) for inviscid barotropic flows corresponds to the universal Scheme (43) with $L = -\varrho\left[\omega + \alpha \overset{\odot}{\beta} + \frac{1}{2}(\alpha\nabla\beta)^2 + e(\varrho)\right]$ but also any other Lagrangian of a continuous system with Galilean invariance and equality of momentum density and mass flux density. Moreover, in [56], it is certified that the velocity field fulfils all transformation rules with respect to the Galilean Group in any case.

For the universal Scheme (43), the mass density (Equation ((37)) simplifies to

$$\varrho = -\frac{\partial L}{\partial \omega}\,, \tag{48}$$

while the momentum density, which equals the mass flux density, according to Equation (39), becomes:

$$\varrho\vec{u} = \vec{p} = \underbrace{-\frac{\partial L}{\partial \omega}}_{\varrho}\nabla\varphi - \frac{\partial L}{\partial \overset{\odot}{\vartheta}_i}\nabla\vartheta_i\,, \tag{49}$$

Therefore, the velocity field takes the form:

$$\vec{u} = \nabla\varphi + \gamma_i\nabla\vartheta_i\,, \tag{50}$$

of a generalised Clebsch representation with:

$$\gamma_i = \gamma_i\left(\omega, \vartheta_j, \overset{\odot}{\vartheta}_j, \nabla\vartheta_j\right) := -\frac{1}{\varrho}\frac{\partial L}{\partial \overset{\odot}{\vartheta}_i}\,. \tag{51}$$

The surprising result from the above investigations is that the Clebsch representation of the velocity field is an inevitable consequence of the Galilean invariance. More precisely, the essence of the above can be formulated in the following theorem:

Theorem 1. *If a continuum, no matter whether it is a solid or a fluid, fulfils the following three requirements:*

- *Its dynamics are deducible from Hamilton's Principle (30),*
- *The associated Lagrangian fulfils Galilean invariance, i.e., invariance with respect to (21–23),*
- *Equivalence $\vec{p} = \varrho\vec{u}$ of momentum density and mass flux density is given,*

then the velocity field takes, via Noether's theorem, the analytic form (Equation (50)) of a (generalised) Clebsch representation.

In terms of the above theorem, the Clebsch representation is substantiated from fundamental symmetries in physics.

2.3. An Extended Clebsch Transformation for Viscous Flow

Consider now the NS equations together with the continuity equation:

$$\frac{\mathrm{D}\vec{u}}{\mathrm{D}t} - \nu\Delta\vec{u} + \nabla\left[\frac{p}{\varrho} + U\right] = \vec{0}\,, \tag{52}$$

$$\nabla\cdot\vec{u} = 0\,, \tag{53}$$

assuming incompressible flow according to Equation (53), commensurate with the classical theory of viscous flow [19], such that ν denotes the kinematic viscosity. As demonstrated in Section 2.1 for inviscid flow, the key feature of the Clebsch transformation is that it enables the Euler equations to be written in the form of Equation (10), being a natural outcome of the mathematical structure of inviscid flow theory. For the viscous case, it is less obvious how the transformation applies, since the specific viscous force $-\nu\Delta\vec{u}$ in the NS equations reads, when written in terms of the Clebsch variables, as:

$$-\nu\Delta\vec{u} = 2\nu\nabla\times\vec{\omega} - 2\nu\nabla\underbrace{(\nabla\cdot\vec{u})}_{0} = \nu\Delta\beta\nabla\alpha - \nu\Delta\alpha\nabla\beta - \nu\left(\nabla\alpha\cdot\nabla\right)\nabla\beta + \nu\left(\nabla\beta\cdot\nabla\right)\nabla\alpha\,, \tag{54}$$

which does not correspond directly to the mathematical Scheme (10). Despite this, it is shown in [60] that it is possible to decompose the viscous term according to Equation (10) by introducing an additional field ξ. In addition and as demonstrated subsequently, this procedure applies to any vector field, not only to the specific viscous friction force (Equation (54)).

Theorem 2. *Let $\vec{a}$ be an arbitrary vector field, $\vec{u}$ the velocity field given in Clebsch representation according to Equation (5) based on the Clebsch variables φ, α, β and $\vec{\omega}$ the vorticity given according to Equation (14). Then, a decomposition of $\vec{a}$ according to Scheme (10) is always possible, i.e., three fields ξ, λ and μ exist fulfilling:*

$$\vec{a} = \nabla\xi + \lambda\nabla\alpha + \mu\nabla\beta\,, \tag{55}$$

for prescribed fields α, β where the auxiliary field ξ results as the solution of the conditional equation:

$$\vec{\omega}\cdot\nabla\xi = \vec{\omega}\cdot\vec{a} \tag{56}$$

and the two coefficients λ, μ explicitly as:

$$\lambda = \frac{\vec{\omega}\times[\vec{a}-\nabla\xi]}{2\vec{\omega}^2}\cdot\nabla\beta\,, \tag{57}$$

$$\mu = -\frac{\vec{\omega}\times[\vec{a}-\nabla\xi]}{2\vec{\omega}^2}\cdot\nabla\alpha\,. \tag{58}$$

Proof. The conditional Equation (56) is obtained by taking the inner product of Equation (55) with $2\vec{\omega} = \nabla\alpha\times\nabla\beta$. Once having solved Equation (56), the identity:

$$\vec{\omega}\times(\vec{\omega}\times[\vec{a}-\nabla\xi]) = \vec{\omega}\underbrace{(\vec{\omega}\cdot[\vec{a}-\nabla\xi])}_{0} - [\vec{a}-\nabla\xi]\,\vec{\omega}^2 = -\vec{\omega}^2\,[\vec{a}-\nabla\xi] \tag{59}$$

follows and therefore:

$$\begin{aligned}\vec{a}-\nabla\xi &= \frac{(\vec{\omega}\times[\vec{a}-\nabla\xi])\times\vec{\omega}}{\vec{\omega}^2} = \frac{(\vec{\omega}\times[\vec{a}-\nabla\xi])\times(\nabla\alpha\times\nabla\beta)}{2\vec{\omega}^2}\\ &= \underbrace{\frac{(\vec{\omega}\times[\vec{a}-\nabla\xi])\cdot\nabla\beta}{2\vec{\omega}^2}}_{=:\lambda}\nabla\alpha \underbrace{-\frac{(\vec{\omega}\times[\vec{a}-\nabla\xi])\cdot\nabla\alpha}{2\vec{\omega}^2}}_{=:\mu}\nabla\beta\end{aligned} \tag{60}$$

as the desired decomposition of $\vec{a}-\nabla\xi$ as a linear combination of $\nabla\alpha$ and $\nabla\beta$ according to Equation (55). □

Like the Clebsch variables Φ, α, β, the auxiliary field ξ is not uniquely given, since any particular solution ξ_{p} of the inhomogeneous linear first order PDE (56) can be superposed with any solution ξ_{h} of the respective homogeneous PDE $\vec{\omega}\cdot\nabla\xi_{\mathrm{h}} = 0$. Since three independent solutions are given by α, β and t, the mathematical theory of linear first order PDE implies $\xi_{\mathrm{h}} = F(\alpha,\beta,t)$ for an arbitrary function F. As a consequence:

$$\xi \longrightarrow \xi' = \xi + F(\alpha,\beta,t), \tag{61}$$

is a gauge transformation for the auxiliary field, which is used subsequently to provide a favourable form of the resulting equations.

Note that the decomposition in Equation (60) is applicable to an arbitrary vector field $\vec{a}$, including for instance non-conservative forces. In the case of viscous flow, the specific viscous force in Equation (54) in the NS Equations (52) can be partitioned into two parts [60], namely $\vec{a} = \nu(\nabla\beta \cdot \nabla)\nabla\alpha - \nu(\nabla\alpha \cdot \nabla)\nabla\beta$ on the one hand and $-\nu\nabla^2\vec{u} - \vec{a} = \nu\Delta\beta\nabla\alpha - \nu\Delta\alpha\nabla\beta$ on the other, which makes sense at first glance since the latter contributions already have the desired form of Equation (10). However, a major disadvantage of this partition is that the selected field $\vec{a}$ is not gauge invariant with respect to the transformation shown in Equation (6). Although not representing a problem mathematically, this feature makes the field equations resulting from the generalised Clebsch transformation less transparent from a physical viewpoint.

In contrast to the above, the full specific viscous force in Equation (54) is considered in the following, i.e.,

$$\vec{a} = 2\nu\nabla \times \vec{\omega}\,, \tag{62}$$

leading to a physically more transparent representation of the field equations, as provided in [60]. This procedure allows the NS Equations (52) to be written as:

$$\vec{0} = \nabla\left[\frac{\partial\varphi}{\partial t} + \alpha\frac{\partial\beta}{\partial t} + \frac{\vec{u}^2}{2} + \frac{p}{\varrho} + U + \xi\right] + \left[\frac{\mathrm{D}\alpha}{\mathrm{D}t} + \mu\right]\nabla\beta + \left[-\frac{\mathrm{D}\beta}{\mathrm{D}t} + \lambda\right]\nabla\alpha\,, \tag{63}$$

with ξ, λ, μ given according to Theorem 2. As in Section 2.1, by proper gauging of the potentials, the bracketed terms in Equation (63) vanish separately, resulting in the following set of PDEs

$$\frac{\partial\varphi}{\partial t} + \alpha\frac{\partial\beta}{\partial t} + \frac{\vec{u}^2}{2} + \frac{p}{\varrho} + U + \xi = 0 \tag{64}$$

$$\frac{\mathrm{D}\alpha}{\mathrm{D}t} - \frac{\vec{\omega} \times [2\nu\nabla \times \vec{\omega} - \nabla\xi]}{2\vec{\omega}^2} \cdot \nabla\alpha = 0 \tag{65}$$

$$\frac{\mathrm{D}\beta}{\mathrm{D}t} - \frac{\vec{\omega} \times [2\nu\nabla \times \vec{\omega} - \nabla\xi]}{2\vec{\omega}^2} \cdot \nabla\beta = 0 \tag{66}$$

supplemented by the conditional equation:

$$\vec{\omega} \cdot \nabla\xi = 2\nu\vec{\omega} \cdot (\nabla \times \vec{\omega}) \tag{67}$$

for the auxiliary field and the continuity Equation (53). The latter, in terms of Clebsch variables [43], becomes:

$$\nabla^2\varphi + \alpha\nabla^2\beta + 2\nabla\alpha \cdot \nabla\beta = 0\,. \tag{68}$$

Physically, Equation (64) can be interpreted as a generalised Bernoulli's equation and Equations (65) and (66) as generalised transport equations covering the entire vortex dynamics of the flow.

2.4. Axisymmetric Stagnation Flow

The problem of an axisymmetric stagnation flow against a solid wall, see Figure 1, is considered one of prototypical character in fluid mechanics, since the underlying features of Navier–Stokes theory are exposed, especially the formation of a boundary layer at the solid wall $z = 0$. It is one of the few boundary layer problems that allow for an analytical treatment of the full NS equations, without the necessity of neglecting higher order terms. In the inviscid case, the velocity field, written in cylindrical coordinates (r, ϕ, z), is given by Mayes et al. [61]:

$$\vec{u}_{\text{invis}} = Ar\vec{e}_r - 2Az\vec{e}_z\,. \tag{69}$$

Although Equation (69) is a solution of the NS equations, it does not match to the no-slip condition $\vec{e}_r \cdot \vec{u} = 0$ at the wall and therefore represents a good approximation for the far field only, i.e., the field in the region $z \gg \sqrt{\nu/A}$ far beyond the boundary layer; whereas, in the vicinity of the wall, a boundary layer becomes manifest in which the velocity is assumed to take the slightly different form:

$$\vec{u} = rf'(z)\vec{e}_r - 2f(z)\vec{e}_z\,, \tag{70}$$

containing the function $f(z)$ that has to be determined. Note that the continuity Equation (53) is fulfilled identically by Equation (70). The associated boundary conditions are (i) the no-slip/no-penetration condition $\vec{u} = \vec{0}$ at $z = 0$ and (ii) the matching condition $\vec{u} \to \vec{u}_{\text{invis}}$ for $z \to \infty$, leading to:

$$f(0) = 0\,, \qquad f'(0) = 0\,, \qquad \lim_{z\to\infty} f'(z) = A\,. \tag{71}$$

Now the extended Clebsch transformation developed in Section 2.3 is applied: writing the vorticity and the specific viscous force as:

$$\vec{\omega} = \frac{r}{2} f''(z)\,\vec{e}_\phi\,, \tag{72}$$

$$2\nu\nabla \times \vec{\omega} = \nu\left[2f''(z)\,\vec{e}_z - rf'''(z)\,\vec{e}_r\right]\,, \tag{73}$$

the conditional equation simplifies to:

$$\frac{\partial \xi}{\partial \phi} = 0\,, \tag{74}$$

implying $\xi = \xi(r,z,t)$ as a general solution. Since a steady axisymmetric flow is considered, it is subsequently assumed independent of time and azimuthal angle for all potentials ξ, α, β and φ. The remaining equations of interest are the generalised transport Equations (65) and (66) which, after minor mathematical manipulation, take the form:

$$\left[r^2 f'(z)\,f''(z) - 2\nu f''(z) + \frac{\partial \xi}{\partial z}\right]\frac{\partial \alpha}{\partial r} - \left[2rf(z)f''(z) + \nu r f'''(z) + \frac{\partial \xi}{\partial r}\right]\frac{\partial \alpha}{\partial z} = 0\,, \tag{75}$$

$$\left[r^2 f'(z)\,f''(z) - 2\nu f''(z) + \frac{\partial \xi}{\partial z}\right]\frac{\partial \beta}{\partial r} - \left[2rf(z)f''(z) + \nu r f'''(z) + \frac{\partial \xi}{\partial r}\right]\frac{\partial \beta}{\partial z} = 0\,. \tag{76}$$

The generalised Bernoulli Equation (64) and the continuity Equation (68) are decoupled from Equations (75) and (76); they provide the third Clebsch variable φ and the pressure p, both of which are not required here.

Three unknown fields ξ, α, β are involved in the two decisive PDEs (75) and (76) due to the freedoms given by the gauge symmetry with respect to the transformation in Equation (6). This provides the opportunity to choose the potentials in a beneficial way: for instance, by considering that the vorticity in Equation (72) can be written as $2\vec{\omega} = rf''(z)\,\vec{e}_\phi = rf''(z)\,\vec{e}_z \times \vec{e}_r = \nabla f'(z) \times \nabla r^2/2$ and comparison with Equation (14) motivates the choice:

$$\alpha = f'(z) \tag{77}$$

$$\beta = \frac{r^2}{2} \tag{78}$$

for the two Clebsch variables. By inserting these into Equations (75) and (76), the following simplified equations:

$$\frac{\partial \xi}{\partial r} = -2rf(z)f''(z) - \nu r f'''(z)\,, \tag{79}$$

$$\frac{\partial \xi}{\partial z} = 2\nu f''(z) - r^2 f'(z)\,f''(z)\,, \tag{80}$$

result for the auxiliary field which are both integrable, leading to:

$$\xi = -r^2 f(z) f''(z) - \frac{\nu}{2} r^2 f'''(z) + F_1(z)\,, \tag{81}$$

$$\xi = 2\nu f'(z) - \frac{r^2}{2} f'(z)^2 + F_2(r)\,, \tag{82}$$

containing the two integration functions $F_1(z)$ and $F_2(r)$. By subtracting Equation (81) from (82), the auxiliary field is eliminated, giving:

$$\frac{r^2}{2}\left[\nu f'''(z) + 2f(z)f''(z) - f'(z)^2\right] + F_2(r) + 2\nu f'(z) - F_1(z) = 0\,. \tag{83}$$

Next, taking the limit $r \to 0$ of the above equation leads to $F_1(z) = F_2(0) + 2\nu f'(z)$; alternatively, the limit $z \to \infty$, in tandem with the matching condition $f'(z) \to A$ according to Equation (71), gives $F_2(r) - F_2(0) = A^2 r^2/2$. On re-inserting these explicit forms of the integration functions into Equation (83), the following ODE for the unknown function $f(z)$ is obtained:

$$\nu f'''(z) + 2f(z)f''(z) - f'(z)^2 + A^2 = 0. \tag{84}$$

Finally, the substitution $f(z) = \sqrt{\nu A}\bar{f}(\bar{z})$ with $\bar{z} = \sqrt{A/\nu} z$ transforms the latter into the more convenient form:

$$\bar{f}'''(\bar{z}) + 2\bar{f}(\bar{z})\bar{f}''(\bar{z}) - \bar{f}'(\bar{z})^2 + 1 = 0\,, \tag{85}$$

well-known in the literature as the *Falkner-Skan equation* [61].

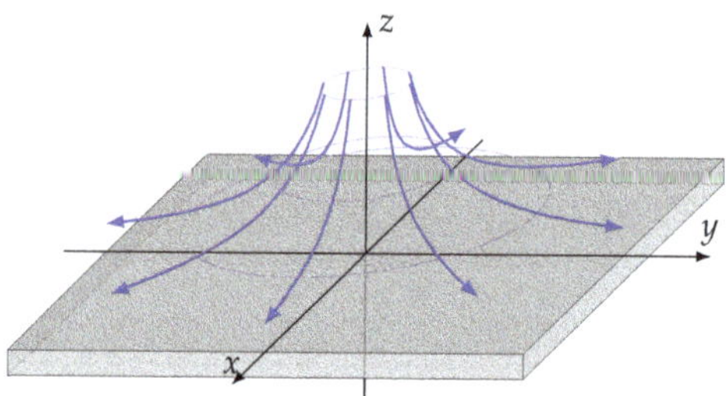

Figure 1. Schematic of an axisymmetric stagnation flow in the vicinity of a solid wall.

3. Complex Variable and Tensor Potential Approach

3.1. The Classical Complex Variable Method

Consider the case of an arbitrary homogeneous continuum, with plane stress given by the tensor:

$$\underline{T} = \begin{pmatrix} \sigma_x & \tau_{xy} \\ \tau_{xy} & \sigma_y \end{pmatrix} \tag{86}$$

under a load provided by a conservative force with specific potential energy U. Assuming a steady state, the equilibrium condition:

$$\nabla \cdot \underline{T} + \varrho \nabla U = \vec{0} \tag{87}$$

has to be fulfilled. By defining the complex variable:

$$\xi := x + \mathrm{i}y\,, \tag{88}$$

the three fields σ_x, σ_y and τ_{xy} can be considered as functions of ξ and its complex conjugate, $\overline{\xi}$, and the equilibrium Condition (87) reads:

$$\vec{0} = \left(\frac{\partial}{\partial\xi} + \frac{\partial}{\partial\overline{\xi}}, \mathrm{i}\frac{\partial}{\partial\xi} - \mathrm{i}\frac{\partial}{\partial\overline{\xi}}\right) \begin{pmatrix} \sigma_x + \varrho U & \tau_{xy} \\ \tau_{xy} & \sigma_y + \varrho U \end{pmatrix}$$
$$= \frac{1}{2} \begin{pmatrix} \frac{\partial}{\partial\overline{\xi}} \left[\sigma_x + \sigma_y + 2\varrho U\right] + \frac{\partial}{\partial\xi} \left[\sigma_x - \sigma_y + \mathrm{i}\tau_{xy}\right] + \text{c.c.} \\ -\mathrm{i}\frac{\partial}{\partial\overline{\xi}} \left[\sigma_x + \sigma_y + 2\varrho U\right] - \mathrm{i}\frac{\partial}{\partial\xi} \left[\sigma_x - \sigma_y + \mathrm{i}\tau_{xy}\right] + \text{c.c.} \end{pmatrix} \tag{89}$$

where c.c. denotes the conjugate complex of the preceding expression. It is convenient to introduce the hydrostatic pressure $p := (\sigma_x + \sigma_y)/2$ and the complex stress $\underline{\sigma} := (\sigma_x - \sigma_y)/2 + \mathrm{i}\tau_{xy}$, allowing Equation (89) to be written as $\Re\left(\partial\left[p + \varrho U\right]/\partial\overline{\xi} + \partial\underline{\sigma}/\partial\xi\right) = 0$ and $\Im\left(\partial\left[p + \varrho U\right]/\partial\overline{\xi} + \partial\underline{\sigma}/\partial\xi\right) = 0$ and therefore:

$$\frac{\partial}{\partial\overline{\xi}} \left[\frac{p}{\varrho} + U\right] + \frac{1}{\varrho}\frac{\partial\underline{\sigma}}{\partial\xi} = 0, \tag{90}$$

as the complex form of the equilibrium Condition (87). The equilibrium Condition (90) is fulfilled identically by introducing a real-valued potential field Φ according to:

$$\frac{\underline{\sigma}}{\varrho} = -4\frac{\partial^2\Phi}{\partial\overline{\xi}^2}, \tag{91}$$

$$\frac{p}{\varrho} + U = 4\frac{\partial^2\Phi}{\partial\overline{\xi}\partial\xi}. \tag{92}$$

The potential Φ is the well known *Airy stress function* from the theory of linear elasticity [32,62]. Since, however, no assumptions regarding the constitutive equations of the respective continuum are required for the above derivation of the complex equilibrium Condition (87) and the use of Airy's stress function according to Equations (91) and (92), this approach applies to any continuum and thus also to Stokes flow. By assuming a steady flow and neglecting the nonlinear terms for the case of very small Reynolds numbers, the NS Equations (52) simplify to the Stokes equation; being of the general form in Equation (87) with stress tensor $T_{ij} = -p\delta_{ij} + \varrho\nu\left(\partial_i u_j + \partial_j u_i\right)$, leading to a complex stress field:

$$\frac{\underline{\sigma}}{\varrho} = -2\nu\frac{\partial u}{\partial\overline{\xi}}, \tag{93}$$

where u denotes the complex velocity:

$$u = u_x + \mathrm{i}u_y\,. \tag{94}$$

On introducing a stream function Ψ, satisfying:

$$u = -2\mathrm{i}\frac{\partial\Psi}{\partial\overline{\xi}}, \tag{95}$$

which fulfils the continuity Equation (53) identically [37,38], the constitutive Equation (93), written in terms of the stream function, becomes:

$$\frac{\underline{\sigma}}{\varrho} = 4\mathrm{i}\nu\frac{\partial^2\Psi}{\partial\overline{\xi}^2}. \tag{96}$$

By inserting Equation (96) into (91), the fully integrable equation:

$$\frac{\partial^2}{\partial\overline{\xi}^2}\left[\Phi + \mathrm{i}\nu\Psi\right] = 0, \tag{97}$$

results; the double integration of which gives the general solution:

$$\Phi + \mathrm{i}\nu\Psi = g_0(\xi) + \bar{\xi} g_1(\xi), \tag{98}$$

containing integration functions $g_0(\xi)$ and $g_1(\xi)$, known as *Goursat functions* [62]. Note that, such an approach leads to the well-known Sherman–Lauricella equations [63–65] and to a reproduction of the Muskhelishvili–Kolosov formula [32,33].

The complex variable method has been applied successfully to various Stokes' flow problems [35,66–69], usually by adopting a conformal mapping on the Goursat functions known for a simple flow domain in order to obtain the respective Goursat functions for a non-trivial domain. Being holomorphic, the Goursat functions can be recovered alternatively from their boundary values determined by the boundary conditions as utilised in [70–72] as an investigative tool for studying the internal flow structure of film flows over corrugated walls.

3.2. Integration of the Full 2D Navier–Stokes Equations

Making use again of the complex variable transformation in Equations (88) and (94), the NS Equations (52) and continuity Equation (53) can be reformulated as:

$$\frac{\partial u}{\partial t} + 2\frac{\partial}{\partial \bar{\xi}}\left[\frac{\bar{u}u}{2} + \frac{p}{\varrho} + U\right] + 2\frac{\partial}{\partial \xi}\left(\frac{u^2}{2}\right) = 4\nu\frac{\partial^2 u}{\partial \bar{\xi}\partial \xi}, \tag{99}$$

$$\Re\left(\frac{\partial u}{\partial \xi}\right) = 0, \tag{100}$$

in terms of the complex coordinate ξ, the complex velocity field u and their complex conjugates, with $\Re$ denoting the real part of the subsequent complex expression. It is obvious that by introducing a stream function Ψ, according to Equation (95), the continuity Equation (100) is fulfilled identically. Accordingly, the complex NS Equation (99) can be written as:

$$\frac{\partial}{\partial \bar{\xi}}\left[\frac{\bar{u}u}{2} + \frac{p}{\varrho} + U - \mathrm{i}\frac{\partial \Psi}{\partial t}\right] + \frac{\partial}{\partial \xi}\left(\frac{u^2}{2}\right) = 2\nu\frac{\partial^2 u}{\partial \bar{\xi}\partial \xi}. \tag{101}$$

By introducing a new complex potential M according to

$$\frac{\bar{u}u}{2} + \frac{p}{\varrho} + U - \mathrm{i}\frac{\partial \Psi}{\partial t} = 2\frac{\partial M}{\partial \xi}, \tag{102}$$

an integrable form of Equation (99),

$$\frac{\partial}{\partial \xi}\left[2\frac{\partial M}{\partial \bar{\xi}} + \frac{u^2}{2} - 2\nu\frac{\partial u}{\partial \bar{\xi}}\right] = 0, \tag{103}$$

is obtained which, after integration with respect to ξ, gives:

$$2\frac{\partial M}{\partial \bar{\xi}} + \frac{u^2}{2} - 2\nu\frac{\partial u}{\partial \bar{\xi}} = f(\bar{\xi}), \tag{104}$$

having integration function $f(\bar{\xi})$ on the right hand side. The latter can be conveniently set to zero by re-gauging the potential M as follows:

$$M \longrightarrow M + \frac{1}{2}F(\bar{\xi}),$$

with $F'(\bar{\xi}) = f(\bar{\xi})$, since according to Equation (102) any complex function of $\bar{\xi}$ can be added to M without having any effect. Making use of Equation (95), Equation (104) simplifies to:

$$2\frac{\partial}{\partial\bar{\xi}}\left[M + 2\mathrm{i}\nu\frac{\partial\Psi}{\partial\bar{\xi}}\right] + \frac{u^2}{2} = 0\,. \tag{105}$$

Finally, a second complex potential χ is introduced via:

$$M + 2\mathrm{i}\nu\frac{\partial\Psi}{\partial\bar{\xi}} = 2\frac{\partial\chi}{\partial\bar{\xi}}\,, \tag{106}$$

by which the two complex equations together with Equations (102) and (105) take the final form:

$$\frac{\bar{u}u}{2} + \frac{p}{\varrho} + U - \mathrm{i}\left[\frac{\partial\Psi}{\partial t} - 4\nu\frac{\partial^2\Psi}{\partial\bar{\xi}\partial\xi}\right] = 4\frac{\partial^2\chi}{\partial\bar{\xi}\partial\xi}, \tag{107}$$

$$-\frac{u^2}{2} = 4\frac{\partial^2\chi}{\partial\bar{\xi}^2}. \tag{108}$$

By Equations (107) and (108), two complex equations are given containing the complex potential χ, the real-valued stream function Ψ and the pressure p as unknowns. They constitute a *first integral of Navier–Stokes equations* [73] in the sense that by taking the difference of the derivative of Equation (108) with respect to ξ with the derivative of Equation (107) with respect to $\bar{\xi}$, the complex NS Equation (99) is recovered.

Compared to the classical complex variable approach outlined in Section 3.1, a complex valued potential field χ is required for the integration of the full 2D-NS equation in place of the real valued Airy stress function Φ utilised for the Stokes equation. A relationship between χ and Φ can be established for the particular case of *steady flow*: by setting $\partial\Psi/\partial t = 0$, Equation (107) simplifies to:

$$\frac{\bar{u}u}{2} + \frac{p}{\varrho} + U = 4\frac{\partial^2}{\partial\bar{\xi}\partial\xi}\left[\chi - \mathrm{i}\nu\Psi\right]\,,$$

which, apart from the additional nonlinear term $\bar{u}u/2$, corresponds to Equation (92) if $\Phi = \chi - \mathrm{i}\nu\Psi$ is identified as the Airy stress function. Thus, for the steady flow case the two field Equations (107) and (108) simplify to:

$$\frac{\bar{u}u}{2} + \frac{p}{\varrho} + U = 4\frac{\partial^2\Phi}{\partial\bar{\xi}\partial\xi}\,, \tag{109}$$

$$4\mathrm{i}\nu\frac{\partial^2\Psi}{\partial\bar{\xi}^2} + \frac{u^2}{2} = -4\frac{\partial^2\Phi}{\partial\bar{\xi}^2}\,, \tag{110}$$

in accordance with [38]. Note that Equation (109) takes the form of Bernoulli's equation apart from the second order derivative of the potential on its right hand side.

3.3. Integration of the Dynamic Boundary Condition

The mathematical derivation is completed by the specification of appropriate boundary conditions, which take the form of no-slip/no-penetration conditions at solid walls, inflow and outflow conditions and, in the case of film or multiphase flows, kinematic and dynamic boundary conditions at a free surface or internal interface. These are discussed in detail in [38,73]. However, a key feature of the above approach that requires emphasising is that the dynamic boundary condition associated with a free surface or internal interface can be similarly integrated in the case of steady flow leading to a corresponding first integral form that greatly simplifies enforcing such a condition compared to the standard method of addressing such problems employing the standard NS equations and boundary conditions written in primitive (i.e., observable) variables. Accordingly, this case as outlined briefly below.

Consider a simply connected domain with a boundary $\xi = f\,(s)$, parametrised with respect to the arc length s of the boundary; furthermore, tangential and normal unit vectors along the boundary are taken to be $f'(s)$ and $n = \mathrm{i}f'(s)$, respectively. In terms of the stream function Ψ, the kinematic boundary condition $\Im(u\bar{f}') = 0$ ensures that along a free surface/interface the velocity in the normal direction vanishes, resulting in:

$$\frac{\partial \Psi}{\partial s} = 0\,. \tag{111}$$

The original dynamic boundary condition, after transformation into complex representation, reads:

$$\frac{p_0 - p}{\varrho} n - 2\mathrm{i}\nu \frac{\partial u}{\partial \bar{\xi}} \bar{f}' = \frac{\sigma}{\varrho} f''\,. \tag{112}$$

By making use of Equations (109) and (110), the pressure p and the viscous stresses $\mathrm{i}\nu\partial u/\partial\bar{\xi}$ can be replaced by derivatives of Φ, implying:

$$-u\frac{\mathrm{d}\Psi}{\mathrm{d}s} + \left(U + \frac{p_0}{\varrho}\right) n = \frac{\mathrm{d}}{\mathrm{d}s}\left[\frac{\sigma}{\varrho} f' + 4\mathrm{i}\frac{\partial \Phi}{\partial \bar{\xi}}\right]\,, \tag{113}$$

as the boundary condition for Φ. Since the specific potential energy U can be gauged with a constant, $U + p_0/\varrho$ can be replaced by U. Making use of the kinematic boundary condition in Equation (111), integration of Equation (113) leads to:

$$4\mathrm{i}\frac{\partial \Phi}{\partial \bar{\xi}} = -\frac{\sigma}{\varrho} f' + \int U n \mathrm{d}s\,, \tag{114}$$

as a first integral of the dynamic boundary condition. In [38], the same boundary condition is derived in real-valued form and where it is demonstrated that it can be decomposed into a Dirichlet and a Neumann boundary condition for Φ. The reduction of the original non-standard boundary condition in Equation (112) containing mixed contributions from different fields u and p to a mathematical standard form has proven to be extremely beneficial for the implementation of numerical methods of solution, as demonstrated in [6,38,74].

3.4. *Particular Flow Geometries as Exemplars*

Three different flow problems are used to demonstrate applications of the tensor approach described above; two of which lead to closed form analytical solutions, the third requiring a numerical solution.

3.4.1. Uniaxial Flow: Flow over an Oscillating Plate

Consider the case of a uniaxial flow geometry with, $u_y = 0$, or equivalently in complex formulation: $u - \bar{u} = 0$. With reference to Equation (95), the following PDE:

$$\frac{\partial \Psi}{\partial \xi} + \frac{\partial \Psi}{\partial \bar{\xi}} = 0$$

has to be satisfied, implying the following explicit forms:

$$\Psi = \Psi\left(\frac{\xi + \bar{\xi}}{2\mathrm{i}}, t\right) = \Psi\,(y, t) \tag{115}$$

$$u = -2\mathrm{i}\frac{\partial \Psi}{\partial \bar{\xi}} = \Psi'\,(y, t) \tag{116}$$

for the stream function and the velocity; the prime denotes differentiation with respect to y. The general solution of Equation (108) is given by:

$$\chi = \bar{\xi} w_0(\xi, t) + w_1(\xi, t) + \frac{1}{2} \int \left[\int \Psi'(y,t)^2 \, dy \right] dy, \tag{117}$$

containing the two analytic functions $w_0(\xi, t)$, $w_1(\xi, t)$, being generalisations of the so-called *Goursat functions* [38]. Following the insertion of Equation (117), Equation (107) simplifies to:

$$\frac{p}{\varrho} + U - \mathrm{i} \left[\dot{\Psi} - \nu \Psi'' \right] = 4 w_0'(\xi, t) :, \tag{118}$$

the dot over a symbol denoting its time-derivative.

A horizontal plate of infinite extent covered by a fluid invokes a flow by a forced oscillatory movement, see Figure 2.

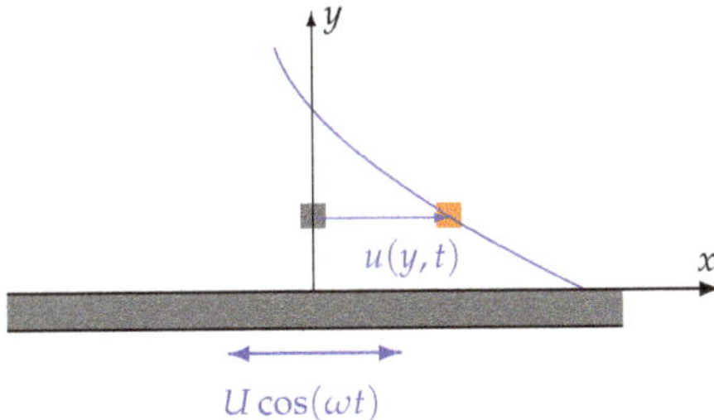

Figure 2. Schematic showing the geometry for the flow over an oscillating plate.

The no-penetration condition is already fulfilled due to the prescribed flow configuration $u_y = 0$, while the no-slip condition compels the fluid to mimic the oscillatory behaviour of the plate according to $u_x(y, 0) = U \cos(\omega t)$. By re-writing the latter in terms of the stream function Ψ via $u_x = \partial \Psi / \partial y$, it takes the form of a Neumann boundary condition $\Psi'(0, t) = U \cos(\omega t)$. Since the stream function is gaugeable by an arbitrary constant, the additional Dirichlet condition $\Psi(0, t) = 0$ can be formulated without loss of generality. Additionally, the asymptotic condition $\Psi'(y, t) \to 0$ has to be fulfilled to ensure that the fluid tends to a state at rest far away from the plate.

Assuming vanishing pressure $p = 0$ and a wave-like solution of the form

$$\Psi = U \Im \left\{ \exp(-\mathrm{i}\omega t) \frac{\exp(\mathrm{i}ky) - 1}{k} \right\}, \tag{119}$$

together with fulfilment of the no-slip and no-penetration condition, then Equation (118) leads to the identity:

$$-\mathrm{i} U \Im \left\{ \frac{-\mathrm{i}\omega + \nu k^2}{k} \exp(\mathrm{i}[ky - \omega t]) + \mathrm{i} \frac{\omega}{k} \exp(-\mathrm{i}\omega t) \right\} = 4 w_0'(\xi, t),$$

requiring k and ω to satisfy the dispersion relation:

$$\mathrm{i}\omega = \nu k^2 \tag{120}$$

for damped transverse waves in accordance to the classical result [22]. The Goursat function w_0 takes the form:

$$w_0 = -\frac{\mathrm{i}}{4} \xi \nu U \Im \left[k \exp(-\mathrm{i}\omega t) \right]. \tag{121}$$

3.4.2. Axisymmetric Flow: The Lamb-Oseen Vortex

Consider a class of flows given by the following particular form:

$$\Psi = \Psi(r,t)\,, \tag{122}$$
$$u = -2\mathrm{i}\frac{\partial\Psi}{\partial\bar{\xi}} = -\frac{\mathrm{i}\xi}{r}\Psi'(r,t)\,, \tag{123}$$
$$r := \sqrt{\bar{\xi}\xi}\,, \tag{124}$$

of the stream function with the prime denoting differentiation with respect to r; that is flows, the streamlines of which form concentric circles, see Figure 3a. Assuming a particular solution for Equation (108) of the form $\chi_\mathrm{P} = \chi_\mathrm{P}(r,t)$ leads to the the following expression:

$$r\frac{\partial}{\partial r}\left(\frac{\chi_\mathrm{P}'}{r}\right) = \frac{1}{2}\Psi'(r,t)^2\,, \tag{125}$$

and therefore:

$$\frac{\chi_\mathrm{P}'}{r} = \frac{1}{2}\int\frac{\Psi'(r,t)^2}{r}\mathrm{d}r\,. \tag{126}$$

By inserting the above into Equation (107), the following simplified PDE results:

$$\frac{p}{\varrho} + U - \mathrm{i}\left[\dot{\Psi} - \frac{\nu}{r}\frac{\partial}{\partial r}\left(r\Psi'\right)\right] - \int\frac{\Psi'(r,t)^2}{r}\mathrm{d}r = 0\,. \tag{127}$$

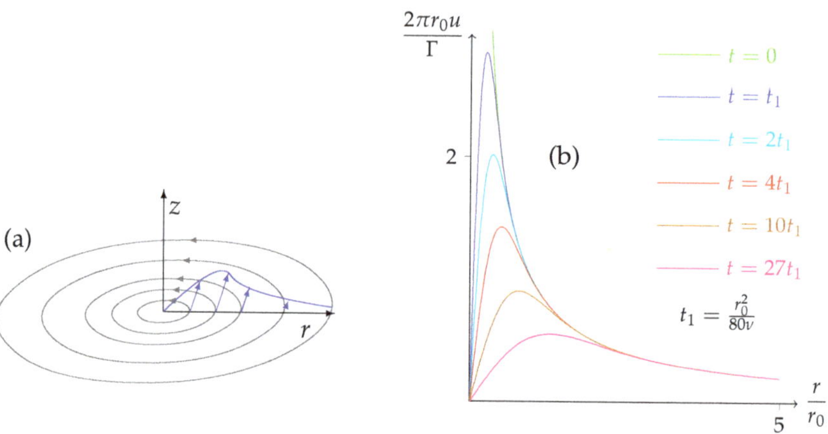

Figure 3. Vortex geometry (**a**) and time evolution of the velocity profile (**b**).

Via the introduction of the following similarity variable:

$$z = \frac{r}{\sqrt{\nu t}}$$

solutions of the form $\Psi = f(z)$ are now searched for. Under the above assumptions the imaginary part of Equation (127) takes the form:

$$\frac{1}{t}\left[\frac{z}{2}f'(z) + \frac{1}{z}\frac{\mathrm{d}}{\mathrm{d}z}\left(zf'(z)\right)\right] = 0$$

which, after making the substitution $g(z) = zf'(z)$, can be written conveniently as;

$$g'(z) + \frac{z}{2}g(z) = 0\,,$$

which, in turn, is easily solved by taking:

$$g(z) = g_0 \exp\left(-\frac{z^2}{4}\right)\,, \tag{128}$$

leading to:

$$f(z) = g_0 \int \frac{1}{z} \exp\left(-\frac{z^2}{4}\right) \mathrm{d}z = \frac{g_0}{2} \mathrm{Ei}\left(-\frac{z^2}{4}\right)\,, \tag{129}$$

where Ei denotes the integral exponential function. This particular solution contains a singularity; however, by considering the superposition:

$$\Psi = \frac{g_0}{2} \mathrm{Ei}\left(-\frac{z^2}{4}\right) - \frac{\Gamma}{2\pi} \ln r,$$

with the well-known potential vortex $\frac{\Gamma}{2\pi} \ln r$ as an alternative solution to Equation (127), the singularity is removed as follows: the complex velocity field resulting from Equation (123) reads

$$u = -\frac{\mathrm{i}\xi}{r} \Psi'(r,t) = \frac{\mathrm{i}}{r}\left[\frac{\Gamma}{2\pi} + g_0 \exp\left(-\frac{r^2}{4\nu t}\right)\right] \exp(\mathrm{i}\varphi)\,,$$

which is convergent in the limit $r \to 0$ if and only if $2\pi g_0 = \Gamma$. Finally, the solution reads:

$$u = \frac{\mathrm{i}\Gamma}{2\pi r}\left[1 - \exp\left(-\frac{r^2}{4\nu t}\right)\right] \exp(\mathrm{i}\varphi)\,, \tag{130}$$

which is a reproduction of the classical Lamb-Oseen vortex [19]. The velocity profile $|u|$ is shown in Figure 3b.

3.4.3. Steady Film Flow over Topography

In the context of the numerical solution of flow problems, possibly involving a free surface, for which inertial effects cannot be ignored, the LSFEM, Cassidy [75], Thatcher [76], Bolton and Thatcher [77], has gained wide acceptability as a very effective and flexible approach—in particular when simple equal order elements in conjunction with highly efficient multigrid solvers are employed [6], exploiting the symmetry and positive definiteness of the resulting system matrices [78].

For reasons of practicality, the complex field Equations (109) and (110) are expressed as real-valued ones in terms of the velocity variables $u_x = \Re u = \partial_y \Psi$, $u_y = \Im u = -\partial_x \Psi$ and first order derivatives, $\phi_x = \partial_x \Phi$, $\phi_y = \partial_y \Phi$, of the Airy stress function leading, together with the continuity Equation (53) and the condition:

$$\frac{\partial \phi_1}{\partial y} - \frac{\partial \phi_2}{\partial x} = 0, \tag{131}$$

to a system of four equations involving first order derivatives of u_x, u_y, ϕ_x, ϕ_y only, conforming ideally to a first-order system least squares methodology.

Figure 4 shows the problem of gravity-driven film flow down a corrugated rigid surface inclined at an angle α to the horizontal considered by Marner [6]. Along the stationary corrugated surface, velocity Dirichlet conditions are imposed. While at the free surface, in addition to a kinematic boundary condition, two dynamic conditions are imposed resulting as inhomogeneous Dirichlet conditions for ϕ_1 and ϕ_2 from Equation (114) by decomposition into real and imaginary parts; these depend on the surface tension, the curvature and the potential energy density. The resulting free surface profile is obtained by iterating over the kinematic condition while solving a sequence of flow problems with prescribed dynamic conditions in a fixed domain using the LSFEM.

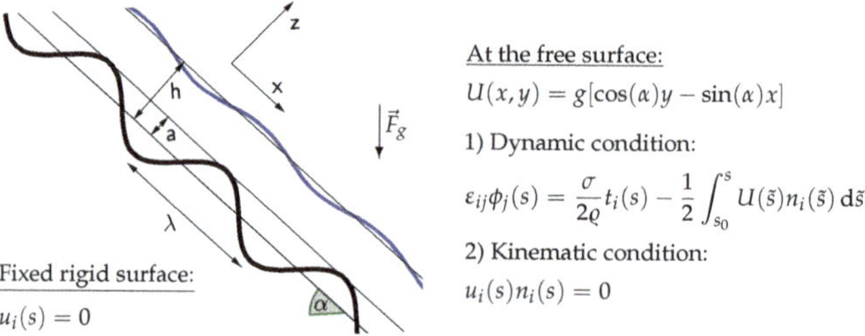

Figure 4. Schematic of gravity-driven film flow down a corrugated rigid surface inclined at an angle α to the horizontal. The indices i and j in the boundary conditions run from 1 to 2.

Two representative results for two differently contoured and repeating surface shapes are shown in Figure 5:

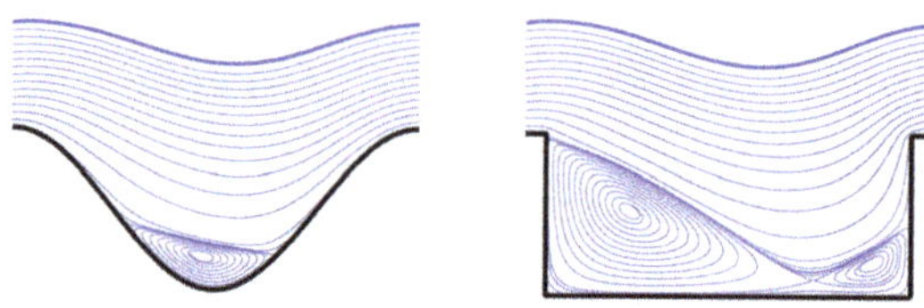

Figure 5. Streamlines elucidating steady film flow over two different periodically corrugated inclined surfaces: on the left a smoothly varying feature, on the right a feature involving sharp, step changes.

The resulting streamline patterns reproduce exactly the observed corresponding experimental results of [79], obtained using silicon oil *Elbesil 145*: $\varrho = 964.8\,\mathrm{kg/m^3}$, $\nu = 144.2\,\mathrm{mm^2/s}$, $\sigma = 20.1\,\mathrm{mN/m}$, $\lambda = 20\,\mathrm{mm}$, $A = 4\,\mathrm{mm}$, $H_0 = 5\,\mathrm{mm}$, $\alpha = 10^\circ$.

Overall, the numerical method was shown [6] to produce an accurate and reliable result over a wide range of Reynolds and capillary numbers for the above problem.

3.5. Tensor Potential Approach

The obvious limitation of the 2D complex-variable formulation is its extension beyond two dimensions, since for the case of 3D viscous flow, a corresponding complex first integral formulation is, by definition, not possible; however, a real-valued tensor form is. In two dimensions, such a form can be established by decomposing both Equations (108) and (107) into real and imaginary parts and taking the linear combinations $\Re(107) \pm \Re(108)$, leading to four real-valued equations:

$$u_1^2 + \frac{p}{\varrho} + U = 2\partial_2^2\Re\chi + 2\partial_1\partial_2\Im\chi\,, \tag{132}$$

$$u_2^2 + \frac{p}{\varrho} + U = 2\partial_1^2\Re\chi - 2\partial_1\partial_2\Im\chi\,, \tag{133}$$

$$u_1u_2 = -\left\{\partial_1^2 - \partial_2^2\right\}\Im\chi - 2\partial_1\partial_2\Re\chi\,, \tag{134}$$

$$\partial_t\Psi = -\nabla^2\left[\Im\chi - \nu\Psi\right], \tag{135}$$

which, in the context of extension to higher dimensions, can be written in the following convenient and compact tensor form:

$$u_i u_j + \left[\frac{p}{\varrho} + U\right] \delta_{ij} = -\partial_k \partial_k \tilde{a}_{ij} + \partial_i A_j + \partial_j A_i - \partial_k A_k \delta_{ij} \,, \tag{136}$$

with the tensor and vector fields given by $\tilde{a}_{ij} = -\Re\chi\delta_{ij}$ and $A_j = \partial_i \tilde{a}_{ij} + \varepsilon_{jk}\partial_k \Im\chi$, respectively. The abbreviations $\partial_k := \partial/\partial x_k$ and $\partial_t := \partial/\partial t$ imply partial differentiation, while δ_{ij} and ε_{ij} denote the Kronecker delta function and the two-dimensional Levi-Civita symbol, respectively.

Recently [39], the corresponding tensor potential formulation for three dimensions has been established, an essential underpinning being analogies drawn with the methodical reduction of Maxwell's equations [6]. The continuity Equation (53) is shown to be fulfilled identically following the introduction a streamfunction vector Ψ_k for the velocity, in accordance with $u_i = \varepsilon_{ijk}\partial_j \Psi_k$ and ε_{ijk} denoting the corresponding 3D Levi-Civita symbol, representing a 3D generalisation of the 2D streamfunction [21]. Compared to the 2D first integral of the NS equations, the corresponding 3D formulation utilises an independent symmetric tensor potential $\tilde{a}_{ij}$ and a vector potential φ_i with the indices i, j taking values from 1 to 3. The auxiliary vector field A_j in the 3D case reads:

$$A_j = \partial_k \tilde{a}_{kj} + \varepsilon_{jlk}\partial_l \varphi_k \,, \tag{137}$$

and Equation (135) has to replaced by its corresponding 3D form:

$$\partial_t \Psi_n = \nu \partial_k \partial_k \Psi_n - \varepsilon_{nkl}\partial_k \partial_m \tilde{a}_{ml} \,. \tag{138}$$

Like the Clebsch variables considered in Section 2, the tensor potential $\tilde{a}_{ij}$ and the vector potential φ_i are not unique and can be gauged in a beneficial way. Obviously, by performing the operations:

$$\tilde{a}_{ij} \longrightarrow \tilde{a}_{ij} + \partial_i \alpha_j + \partial_j \alpha_i - \partial_k \alpha_k \delta_{ij} \,, \tag{139}$$

$$\varphi_i \longrightarrow \varphi_i + \partial_i \zeta \,, \tag{140}$$

for an arbitrary vector field α_i and an arbitrary scalar field ζ, the field Equations (136) remain invariant. These rules are utilised in [39] to establish bona fide gauging scenarios. In the same paper the prescription of appropriate commonly occurring physical and necessary auxiliary boundary conditions, incorporating for completeness the derivation of a first integral of the dynamic boundary condition at a free surface, is established, together with how the general approach can be advantageously reformulated for application in solving unsteady flow problems with periodic boundaries.

Using a tensor formulation, the approach is suitable for use in the case of an arbitrary number of dimensions: in [80], a potential-based first integral form is established for the 4D energy-momentum equations for flows under relativistic conditions.

4. Discussion

Based on a detailed analysis and discourse, the two different approaches considered above can be explained in the light of their different origins: the Clebsch representation of the velocity can, according to recent analysis [56], be understood as a natural outcome of Galilean invariance; whereas, Airy's stress function originates historically from the 2D static equilibrium of internal forces. In the course of a long and growing series of research contributions both approaches have been generalised and made available for use in solving arbitrary flow problems. A Clebsch transformation has emerged that applies to arbitrary forces in the equations of motion, including viscous ones, while extensions to the Airy stress function approach applies to cases beyond static equilibrium, in particular to unsteady flows with inertia and, after rearrangement of the complex equations, to tensor equations (in terms of a tensor potential) for 3D viscous flows and even for the case of 4D relativistic flows. The use of a tensor potential has parallels to Maxwell's theory of electromagnetism [6,39,80]. The capabilities of both approaches have been convincingly demonstrated through the solution of a variety of illustrative examples.

Despite the very positive stage of development of both methods, some open questions remain: the first is that the field Equations (64)–(68) resulting, via the extended Clebsch transformation according to Theorem 2, from the NS equations are not self-adjoint; as in the inviscid flow case. The non-existence of a Lagrangian appears to be due to energy dissipation caused by viscosity. Anthony [81] poses a possible strategy to overcome this problem by including thermal degrees of freedom and related inner energy in order to remain consistent with Noether's theorem, which implies conservation of energy for Lagrangians being invariant with respect to time-translations. Note that the present work, by *dissipation*, the irreversible transfer of energy from mechanical to thermal energy is understood from the physicist's viewpoint, while the total energy (the sum of mechanical and thermal energy) is conserved. Based on Seliger and Witham [24]'s classical work, Zuckerwar and Ash [82,83] suggest a Lagrangian considering only volume viscosity, leading to equations of motion containing, only qualitatively, the effect of volume viscosity but differing quantitatively from the compressible NS equations—also known as the Navier–Stokes–Duhem equations [84,85] without shear viscosity. They interpret their result as a generalisation of the theory of viscous flow towards thermodynamic non-equilibrium. A Lagrangian for viscous flow considering both shear viscosity and volume viscosity has been suggested by Scholle and Marner [86], Marner et al. [87], again leading to equations of motion that differ from the Navier–Stokes–Duhem equations by non-equilibrium terms. A striking feature of their Lagrangian is a *discontinuity*, causing fluctuations on a microscopic scale and revealing parallels to a stochastic variational description as in statistical physics; see, for example, [88–93]. Since these considerations go beyond the scope of classical fluid mechanics, further research on this particular field is required.

A second unanswered question is whether a general and all-encompassing potential approach exists reducing to both the Clebsch and the tensor potential approach as special cases. The search for this 'missing link' between two conceptually different approaches represents another future research topic of general interest. A promising first step could be an analysis of the Clebsch transformation in the sense of general relativity [94], followed by a particular classical limit as shown in Lightman et al. [95], problem 5.31 pp. 35, 227–228.

Thirdly, the first integral of the 4D energy-momentum equations based on a tensor potential [80] points to the future use of mathematical techniques and methods of solutions not currently applicable to the field equations in their original form, in particular the use of matrix structures within the framework of Clifford algebra, based on quaternions or Dirac matrices with the goal of developing highly efficient methods of solution. Having mapped the entire problem to a matrix-algebra framework, the limit $c \to \infty$ could be applied in order to provide efficient solutions of the classical NS equations. Although implementation of such a matrix-algebra techniques remains speculative at this stage, it deserves further investigation since its utilisation can lead to significant economic gains in the computation of fluid flows, in a similar fashion to the use of quaternions representing spatial rotation operations [96,97], potentially leading to the formulation of highly efficient and predictive CFD software.

Author Contributions: Conceptualisation, M.S. and P.H.G.; methodology, M.S. and F.M.; software, F.M.; investigation, F.M. and P.H.G.; writing—original draft preparation, M.S.; writing—review and editing, P.H.G. and F.M.; visualisation, M.S. and F.M.; funding acquisition, M.S., P.H.G. and F.M. All authors have read and agreed to the published version of the manuscript.

Funding: This research was funded by the German Research Foundation (DFG) grant numbers SCHO 767/6-1 and SCHO 767/6-3.

Conflicts of Interest: The authors declare no conflict of interest.

Abbreviations

The following abbreviations are used in this manuscript:

2/3/4D	two-/three-/four-dimensional
LSFEM	least square finite element method
NS	Navier–Stokes
ODE	ordinary differential equation
PDE	partial differential equation

Appendix A. Proof of the Existence of a Representation with Two Pairs of Clebsch Variables

Lin [44,46] proposed a variational principle for superfluid helium resulting by variation in a Clebsch representation with three pairs of variables:

$$\vec{u} = \nabla\Phi + \lambda_1\nabla\xi_1 + \lambda_2\nabla\xi_2 + \lambda_3\nabla\xi_3\,, \tag{A1}$$

where the three fields ξ_i $(i = 1,2,3)$ are identified physically as material coordinates while the three conjugated fields λ_i have been introduced as Lagrange multipliers. It can be proven easily that for a fluid flow a representation of the velocity in the form of Equation (A1) always exist, even if $\Phi = 0$ is assumed. The essence is the existence of both a field representation $\psi = \psi\left(x_i, t\right)$ and material representation $\psi = \Psi\left(\xi_j, t\right)$ for an arbitrary field ψ and a related invertible transformation:

$$x_i = x_i\left(\xi_j, t\right)\,, \tag{A2}$$

between both [22,98]. By taking the gradient of Equation (A2), the relation:

$$\vec{e}_i = \nabla x_i = \frac{\partial x_i}{\partial \xi_j}\nabla\xi_j, \tag{A3}$$

is obtained with the deformation gradient $\frac{\partial x_i}{\partial \xi_j}$. Thus for an arbitrary field $\vec{u} = u_i\vec{e}_i$ it follows that:

$$\vec{u} = u_i\vec{e}_i = \underbrace{u_i\frac{\partial x_i}{\partial \xi_j}}_{\lambda_j}\nabla\xi_j, \tag{A4}$$

corresponding to the form (A1) with $\Phi = 0$.

Next the material representation $\lambda_3 = \Lambda_3\left(\xi_j, t\right)$ for the variable λ_3 is utilised in order to derive the identity:

$$\begin{aligned}\vec{u} &= \lambda_1\nabla\xi_1 + \lambda_2\nabla\xi_2 + \Lambda_3\left(\xi_j, t\right)\nabla\xi_3\\ &= \nabla\int\Lambda_3\left(\xi_j, t\right)\mathrm{d}\xi_3 + \left[\lambda_1 - \frac{\partial}{\partial\xi_1}\int\Lambda_3\left(\xi_j, t\right)\mathrm{d}\xi_3\right]\nabla\xi_1 + \left[\lambda_2 - \frac{\partial}{\partial\xi_2}\int\Lambda_3\left(\xi_j, t\right)\mathrm{d}\xi_3\right]\nabla\xi_2\,,\end{aligned}$$

corresponding via the definitions:

$$\varphi := \int\Lambda_3\left(\xi_j, t\right)\mathrm{d}\xi_3\bigg|_{\xi_j = \xi_j(x_i, t)} \tag{A5}$$

$$\alpha_1 := \lambda_1 - \frac{\partial}{\partial\xi_1}\int\Lambda_3\left(\xi_j, t\right)\mathrm{d}\xi_3\bigg|_{\xi_j = \xi_j(x_i, t)} \tag{A6}$$

$$\alpha_2 := \lambda_2 - \frac{\partial}{\partial\xi_2}\int\Lambda_3\left(\xi_j, t\right)\mathrm{d}\xi_3\bigg|_{\xi_j = \xi_j(x_i, t)} \tag{A7}$$

$$\beta_1 := \xi_1 \tag{A8}$$

$$\beta_2 := \xi_2 \tag{A9}$$

to a Clebsch representation with two pairs of Clebsch variables:

$$\vec{u} = \nabla\varphi + \alpha_1\nabla\beta_1 + \alpha_2\nabla\beta_2 \tag{A10}$$

References

1. Jackson, J.D. *Classical Electrodynamics*, 3rd ed.; Wiley: New York, NY, USA, 1999.
2. Heaviside, O. *Electrical Papers (2 Volumes, Collected Works)*; The Electrician Printing and Publishing Co.: London, UK, 1892.
3. Heaviside, O. *Electromagnetic Theory*; The Electrician Printing and Publishing Co.: London, UK, 1894; Volume 1.
4. Wu, A.; Yang, C.N. Evolution of the concept of the vector potential in the description of fundamental interactions. *Int. J. Mod. Phys. A* **2006**, *21*, 3235–3277. [CrossRef]
5. Jackson, J.D.; Okun, L.B. Historical roots of gauge invariance. *Rev. Mod. Phys.* **2001**, *73*, 663–680. [CrossRef]
6. Marner, F. Potential-Based Formulations of the Navier-Stokes Equations and Their Application. Ph.D. Thesis, Durham University, Durham, UK, 2019.
7. Kaku, M. *Quantum Field Theory: A Modern Introduction*; Oxford University Press: Oxford, UK, 1993.
8. Lanczos, C. The Splitting of the Riemann Tensor. *Rev. Mod. Phys.* **1962**, *34*, 379–389. [CrossRef]
9. Roberts, M.D. The Lanczos Potential for Bianchi Spacetime. *arXiv* **2019**, arXiv:1910.00416.
10. Ehrenberg, W.; Siday, R.E. The Refractive Index in Electron Optics and the Principles of Dynamics. *Proc. Phys. Soc. B* **1949**, *62*, 8–21. [CrossRef]
11. Aharonov, Y.; Bohm, D. Significance of Electromagnetic Potentials in the Quantum Theory. *Phys. Rev.* **1959**, *115*, 485–491. [CrossRef]
12. Boyer, T.H. Does the Aharonov–Bohm Effect Exist? *Found. Phys.* **2000**, *30*, 893–905. [CrossRef]
13. Boyer, T.H. Comment on Experiments Related to the Aharonov–Bohm Phase Shift. *Found. Phys.* **2008**, *38*, 498–505. [CrossRef]
14. Vaidman, L. Role of potentials in the Aharonov-Bohm effect. *Phys. Rev. A* **2012**, *86*, 040101. [CrossRef]
15. Aharonov, Y.; Cohen, E.; Rohrlich, D. Comment on "Role of potentials in the Aharonov-Bohm effect". *Phys. Rev. A* **2015**, *92*. [CrossRef]
16. Vaidman, L. Reply to "Comment on 'Role of potentials in the Aharonov-Bohm effect"'. *Phys. Rev. A* **2015**, *92*, 026102. [CrossRef]
17. Aharonov, Y.; Cohen, E.; Rohrlich, D. Nonlocality of the Aharonov-Bohm effect. *Phys. Rev. A* **2016**, *93*, 042110. [CrossRef]
18. De Oliveira, C.R.; Romano, R.G. A New Version of the Aharonov-Bohm Effect. *Found. Phys.* **2020**, *50*, 137–146. [CrossRef]
19. Lamb, H. *Hydrodynamics*; Cambridge University Press: Cambridge, UK, 1974.
20. Panton, R.L. *Incompressible Flow*; John Wiley & Sons, Inc.: Hoboken, NJ, USA, 1996.
21. Batchelor, G.K. *An Introduction to Fluid Dynamics*; Cambridge Mathematical Library, Cambridge University Press: Cambridge, UK, 2000.
22. Spurk, J.H.; Aksel, N. *Fluid Mechanics*, 2nd ed.; Springer: Berlin/Heidelberg, Germany, 2008. [CrossRef]
23. Clebsch, A. Ueber die Integration der hydrodynamischen Gleichungen. *Journal für die Reine und Angewandte Mathematik* **1859**, *56*, 1–10.
24. Seliger, R.; Witham, G.B. Variational principles in continuum mechanics. *Proc. R. Soc. Lond. A Math. Phys. Eng. Sci.* **1968**, *305*, 1–25. [CrossRef]
25. Wagner, H.J. Das Inverse Problem der Lagrangeschen Feldtheorie in Hydrodynamik, Plasmaphysik und Hydrodynamischem Bild der Quantenmechanik. Ph.D. Thesis, University of Paderborn, Paderborn, Germany, 1997.
26. Calkin, M.G. An action principle for magnetohydrodynamics. *Can. J. Phys.* **1963**, *41*, 2241–2251. [CrossRef]
27. Rund, H. Clebsch representations and relativistic dynamical systems. *Arch. Ration. Mech. Anal.* **1979**, *71*, 199–220. [CrossRef]
28. Madelung, E. Quantentheorie in hydrodynamischer Form. *Zeitschrift für Physik* **1927**, *40*, 322–326. [CrossRef]
29. Schoenberg, M. Vortex Motions of the Madelung Fluid. *Nuovo Cimento* **1955**, *1*, 543–580. [CrossRef]
30. Roberts, M. A fluid generalization of membranes. *Open Phys.* **2011**, *9*, 1016–1021. [CrossRef]

31. Asenjo, F.A.; Mahajan, S.M. Relativistic quantum vorticity of the quadratic form of the Dirac equation. *Phys. Scr.* **2015**, *90*, 015001. [CrossRef]
32. Muskhelishvili, N.I. *Some Basic Problems of the Mathematical Theory of Elasticity*; Noordhoff: Groningen, The Netherlands, 1953.
33. Mikhlin, S.G. *Integral Equations and Their Applications to Certain Problems in Mechanics, Mathematical Physics and Technology*; Pergamon Press: New York, NY, USA, 1957.
34. Legendre, R. Solutions plus complète du problème Blasius. *Comptes Rendus Hebdomadaires des Seances de l Academie des Sciences* **1949**, *228*, 2008–2010.
35. Coleman, C.J. On the use of complex variables in the analysis of flows of an elastic fluid. *J. Non-Newtonian Fluid Mech.* **1984**, *15*, 227–238. [CrossRef]
36. Ranger, K.B. Parametrization of general solutions for the Navier-Stokes equations. *Q. J. Appl. Math.* **1994**, *52*, 335–341. [CrossRef]
37. Scholle, M.; Haas, A.; Gaskell, P.H. A first integral of Navier-Stokes equations and its applications. *Proc. R. Soc. A* **2011**, *467*, 127–143. [CrossRef]
38. Marner, F.; Gaskell, P.H.; Scholle, M. On a potential-velocity formulation of Navier-Stokes equations. *Phys. Mesomech.* **2014**, *17*, 341–348. [CrossRef]
39. Scholle, M.; Gaskell, P.H.; Marner, F. Exact integration of the unsteady incompressible Navier-Stokes equations, gauge criteria, and applications. *J. Math. Phys.* **2018**, *59*, 043101. [CrossRef]
40. Neuber, H. Ein neuer Ansatz zur Lösung räumlicher Probleme der Elastizitätstheorie. *J. Appl. Math. Mech.* **1934**, *14*, 2008–2010.
41. Lee, S.; Ryi, S.; Lim, H. About vortex equations of two dimensional flows. *Indian J. Phys.* **2017**, *91*, 1089–1094. [CrossRef]
42. Greengard, L.; Jiang, S. A New Mixed Potential Representation for Unsteady, Incompressible Flow. *SIAM Rev.* **2019**, *61*, 733–755. [CrossRef]
43. Prakash, J.; Lavrenteva, O.M.; Nir, A. Application of Clebsch variables to fluid-body interaction in presence of non-uniform vorticity. *Phys. Fluids* **2014**, *26*, 077102. [CrossRef]
44. Lin, C.C. Hydrodynamics of Liquid Helium II. *Phys. Rev. Lett.* **1959**, *2*, 245–246. [CrossRef]
45. Eckart, C. Variation Principles of Hydrodynamics. *Phys. Fluids* **1960**, *3*, 421–427. [CrossRef]
46. Lin, C.C. Hydrodynamics of Helium II. In Proceedings of the International School of Physics of Physics "Enrico Fermi", Varenna, Italy, 19–31 August 1963; Academic Press: New York, NY, USA, 1963; Volume 21.
47. Moreau, J.J. Constantes d'un îlot tourbillonnaire en fluide parfait barotrope. *Comptes Rendus Hebdomadaires des Séances de l'Académie des Sciences* **1961**, *252*, 2810–2812.
48. Moffatt, H.K. The degree of knottedness of tangled vortex lines. *J. Fluid Mech.* **1969**, *35*, 117–129. [CrossRef]
49. Yahalom, A. Using Fluid Variational Variables to Obtain New Analytic Solutions of Self-Gravitating Flows with Nonzero Helicity. *Procedia IUTAM* **2013**, *7*, 223–232. [CrossRef]
50. Yahalom, A.; Lynden-Bell, D. Variational principles for topological barotropic fluid dynamics. *Geophys. Astrophys. Fluid Dyn.* **2014**, *108*, 667–685. [CrossRef]
51. Balkovsky, E. Some notes on the Clebsch representation for incompressible fluids. *Phys. Lett. A* **1994**, *186*, 135–136. [CrossRef]
52. Yoshida, Z. Clebsch parameterization: Basic properties and remarks on its applications. *J. Math. Phys.* **2009**, *50*, 113101. [CrossRef]
53. Ohkitani, K.; Constantin, P. Numerical study on the Eulerian–Lagrangian analysis of Navier–Stokes turbulence. *Phys. Fluids* **2008**, *20*, 075102. [CrossRef]
54. Cartes, C.; Bustamante, M.D.; Brachet, M.E. Generalized Eulerian-Lagrangian description of Navier-Stokes dynamics. *Phys. Fluids* **2007**, *19*, 077101. [CrossRef]
55. Ohkitani, K. Study of the 3D Euler equations using Clebsch potentials: dual mechanisms for geometric depletion. *Nonlinearity* **2018**, *31*, R25. [CrossRef]
56. Scholle, M. Construction of Lagrangians in continuum theories. *Proc. R. Soc. Lond. A* **2004**, *460*, 3241–3260. [CrossRef]
57. Schmutzer, E. *Symmetrien und Erhaltungssätze der Physik*; 75: Reihe Mathematik und Physik; Akademie-Verlag: Berlin, Germany, 1972; 165p.
58. Corson, E.M. *Introduction to Tensors, Spinors and Relativistic Wave-Equations: Relation Structure*; Hafner: New York, NY, USA, 1953.

59. Noether, E. Invariant variation problems. *Transp. Theory Stat. Phys.* **1971**, *1*, 186–207. [CrossRef]
60. Scholle, M.; Marner, F. A generalized Clebsch transformation leading to a first integral of Navier-Stokes equations. *Phys. Lett. A* **2016**, *380*, 3258–3261. [CrossRef]
61. Mayes, C.; Schlichting, H.; Krause, E.; Oertel, H.; Gersten, K. *Boundary-Layer Theory*; Physic and Astronomy; Springer: Berlin/Heidelberg, Germany, 2003.
62. Mikhlin, S.; Morozov, N.; Paukshto, M.; Gajewski, H. *The Integral Equations of the Theory of Elasticity*; Teubner-Texte zur Mathematik, Vieweg+Teubner Verlag: Berlin, Germany, 2013.
63. Lauricella, G. Sur l'intégration de l'équation relative àl'équilibre des plaques élastiques encastrées. *Acta Math.* **1909**, *32*, 201. [CrossRef]
64. Sherman, D.I. On the solution of the theory of elasticity static plane problem under given external loading. *Doklady Akademii Nauk SSSR* **1940**, *26*, 25–28.
65. Greengard, L.; Kropinski, M.C.; Mayo, A. Integral equation methods for Stokes flow and isotropic elasticity in the plane. *J. Comput. Phys.* **1996**, *125*, 403–404. [CrossRef]
66. Richardson, S. On the no-slip boundary condition. *J. Fluid Mech.* **1973**, *59*, 707–719. [CrossRef]
67. Howison, S.D. Complex variable methods in Hele–Shaw moving boundary problems. *Eur. J. Appl. Math.* **1992**, *3*, 209–224. [CrossRef]
68. Siegel, M. Cusp formation for time-evolving bubbles in two-dimensional Stokes flow. *J. Fluid Mech.* **2000**, *412*, 227–257. [CrossRef]
69. Cummings, L.J. Steady solutions for bubbles in dipole-driven Stokes flows. *Phys. Fluids* **2000**, *12*, 2162–2168. [CrossRef]
70. Scholle, M.; Wierschem, A.; Aksel, N. Creeping films with vortices over strongly undulated bottoms. *Acta Mech.* **2004**, *168*, 167–193. [CrossRef]
71. Scholle, M. Creeping Couette flow over an undulated plate. *Arch. Appl. Mech.* **2004**, *73*, 823–840. [CrossRef]
72. Scholle, M.; Haas, A.; Aksel, N.; Wilson, M.C.T.; Thompson, H.M.; Gaskell, P.H. Competing geometric and inertial effects on local flow structure in thick gravity-driven fluid films. *Phys. Fluids* **2008**, *20*, 123101. [CrossRef]
73. Marner, F.; Gaskell, P.H.; Scholle, M. A complex-valued first integral of Navier-Stokes equations: Unsteady Couette flow in a corrugated channel system. *J. Math. Phys.* **2017**, *58*, 043102. [CrossRef]
74. Scholle, M.; Gaskell, P.H.; Marner, F. A Potential Field Description for Gravity-Driven Film Flow over Piece-Wise Planar Topography. *Fluids* **2019**, *4*. [CrossRef]
75. Cassidy, M. A Spectral Method for Viscoelastic Extrudate Swell. Ph.D. Thesis, University of Wales, Aberystwyth, UK, 1996.
76. Thatcher, R.W. A least squares method for Stokes flow based on stress and stream functions. In *Manchester Centre for Computational Mathematics*; Report 330; University of Manchester: Manchester, UK, 1998.
77. Bolton, P.; Thatcher, R.W. A least-squares finite element method for the Navier-Stokes equations. *J. Comput. Phys.* **2006**, *213*, 174–183. [CrossRef]
78. Bochev, P.B.; Gunzburger, M.D. *Least-Squares Finite Element Methods*; Applied Mathematical Sciences; Springer: New York, NY, USA, 2009; Volume 166.
79. Schörner, M.; Reck, D.; Aksel, N. Does the topography's specific shape matter in general for the stability of film flows? *Phys. Fluids* **2015**, *27*, 042103. [CrossRef]
80. Scholle, M.; Marner, F.; Gaskell, P.H. A first integral form of the energy-momentum equations for viscous flow, with comparisons drawn to classical fluid flow theory. *Eur. J. Mech. B Fluids* **2020**, under review.
81. Anthony, K.H. Hamilton's action principle and thermodynamics of irreversible processes—A unifying procedure for reversible and irreversible processes. *J. Non-Newton. Fluid Mech.* **2001**, *96*, 291–339. [CrossRef]
82. Zuckerwar, A.J.; Ash, R.L. Variational approach to the volume viscosity of fluids. *Phys. Fluids* **2006**, *18*, 047101. [CrossRef]
83. Zuckerwar, A.J.; Ash, R.L. Volume viscosity in fluids with multiple dissipative processes. *Phys. Fluids* **2009**, *21*, 033105. [CrossRef]
84. Olsson, P. *Transport Phenomena in Newtonian Fluids—A Concise Primer*; Springer Briefs in Applied Sciences and Technology; Springer International Publishing: Cham, Switzerland, 2013. [CrossRef]
85. Belevich, M. *Classical Fluid Mechanics*; Bentham Science Publishers: Sharjah, UAE, 2017.
86. Scholle, M.; Marner, F. A non-conventional discontinuous Lagrangian for viscous flow. *R. Soc. Open Sci.* **2017**, *4*. [CrossRef] [PubMed]

87. Marner, F.; Scholle, M.; Herrmann, D.; Gaskell, P.H. Competing Lagrangians for incompressible and compressible viscous flow. *R. Soc. Open Sci.* **2019**, *6*, 181595. [CrossRef]
88. Cipriano, F.; Cruzeiro, A.B. Navier-Stokes Equation and Diffusions on the Group of Homeomorphisms of the Torus. *Commun. Math. Phys.* **2007**, *275*, 255–269. [CrossRef]
89. Arnaudon, M.; Cruzeiro, A.B.; Galamba, N. Lagrangian Navier-Stokes flows: A stochastic model. *J. Phys. A* **2011**, *44*, 175501. [CrossRef]
90. Arnaudon, M.; Cruzeiro, A.B. Lagrangian Navier-Stokes diffusions on manifolds: variational principle and stability. *Bulletin des Sciences Mathématiques* **2012**, *136*, 857–881. [CrossRef]
91. Arnaudon, M.; Cruzeiro, A.B. Stochastic Lagrangian Flows and the Navier-Stokes Equations. In *Stochastic Analysis: A Series of Lectures*; Springer: Berlin, Germany, 2015; pp. 55–75.
92. Arnaudon, M.; Chen, X.; Cruzeiro, A.B. Stochastic Euler-Poincaré reduction. *J. Math. Phys.* **2014**, *55*, 081507. [CrossRef]
93. Chen, X.; Cruzeiro, A.B.; Ratiu, T.S. Constrained and stochastic variational principles for dissipative equations with advected quantities. *arXiv* **2015**, arXiv:1506.05024.
94. Roberts, M.D. The Clebsch potential approach to fluid Lagrangians. *J. Geom. Phys.* **2017**, *117*, 60–67. [CrossRef]
95. Lightman, A.; Press, W.; Price, R.; Teukolsky, S. *Problem Book in Relativity and Gravitation*; Princeton University Press: Princeton, NJ, USA, 1975.
96. Kuipers, J.B. *Quaternions and Rotation Sequences : A Primer with Applications to Orbits, Aerospace, and Virtual Reality*; : Princeton University Press: Princeton, NJ, USA, 1999.
97. Arribas, M.; Elipe, A.; Palacios, M. Quaternions and the rotation of a rigid body. *Celest. Mech. Dyn. Astron.* **2006**, *96*, 239–251. [CrossRef]
98. Haupt, P. Continuum Mechanics and Theory of Materials. *Appl. Mech. Rev.* **2002**, *55*, B23–B23. [CrossRef]

Article

Stochastic Approaches to Deterministic Fluid Dynamics: A Selective Review

Ana Bela Cruzeiro

Departamento Matemática I.S.T. and Grupo de Física-Matemática University, Av. Rovisco Pais, 1049-001 Lisbon, Portugal; ana.cruzeiro@tecnico.ulisboa.pt

Received: 20 January 2020; Accepted: 13 March 2020; Published: 19 March 2020

Abstract: We present a stochastic Lagrangian view of fluid dynamics. The velocity solving the deterministic Navier–Stokes equation is regarded as a mean time derivative taken over stochastic Lagrangian paths and the equations of motion are critical points of an associated stochastic action functional involving the kinetic energy computed over random paths. Thus the deterministic Navier–Stokes equation is obtained via a variational principle. The pressure can be regarded as a Lagrange multiplier. The approach is based on Itô's stochastic calculus. Different related probabilistic methods to study the Navier–Stokes equation are discussed. We also consider Navier–Stokes equations perturbed by random terms, which we derive by means of a variational principle.

Keywords: Navier–Stokes equation; stochastic Lagrangian flows; stochastic variational principles; stochastic geometric mechanics

---

1. Introduction

The dynamics of an incompressible viscous fluid is modeled by the Navier–Stokes equation, a second order, nonlinear, partial differential equation describing the balance of mass and momentum of the fluid flow. In the absence of external forces and considering a perfectly incompressible fluid, the Navier–Stokes equation reads

$$\frac{\partial}{\partial t}u + (u.\nabla)u = \nu\Delta u - \nabla p \tag{1}$$

where the velocity field u is required to satisfy the incompressibility condition $\mathrm{div}u = 0$, the fluid density being equal to 1. The constant ν denotes the kinematic viscosity and u the fluid velocity. Moreover, the symbol p stands for the pressure within the fluid, and is yet another unknown in the equation. Equation (1) has a huge number of applications in physics and engineering.

The Navier–Stokes equation is deterministic, but it is well known that some of its solutions seem to exhibit random behavior, which might eventually provide an insight into the onset of turbulence although from a purely mathematical point of view, the very important problem of existence and smoothness of the solutions to Equation (1) remains largely unsolved to this day. In this article we will review various ways of introducing randomness into the analysis of the solutions to Equation (1).

In Physics, the most famous deterministic equation hiding randomness is the Schrödinger equation. There is no mathematical probability theory behind it, but there is manifestly randomness in any Quantum Physics Lab. This puzzling situation did not prevent Quantum Theory to develop an impressive corpus of techniques to control, in particular, the transition from (in principle differentiable) solutions of classical equations of motion to some forms of randomness. Is it possible to draw an analogy between this classical/quantum relation and the one between Euler and Navier–Stokes equations? Of course the status of the two hydrodynamical equations are quite different from the above ones. Euler equation, although modeling only "dry water" (Von Neumann) is already quite complicated. It does not seem, yet, to have proof that it does not produce singularities. But the abovementioned analogy could, in particular, provide an insight into the onset of turbulence.

There are different ways to introduce randomness: uncertainty may have its origin in errors in the initial conditions, for example. In this case statistical approaches are considered: one studies the time evolution of some probability measure, supported on the relevant physical initial data. This is part of the statistical approach to turbulence, initiated already in the 19th century (c.f, among many others, [1,2]). Other various types of Langevin dynamics, using stochastic diffusion processes, have been proposed to describe equilibrium and non-equilibrium dynamics as well as Kraichnan's model in turbulent advection (c.f, for instance, [3]). On the other hand uncertainty may be generated by the chosen numerical model (we refer to [4] for a discussion on these issues in climate modeling, see also [5]).

Another very popular way to introduce stochasticity is to perturb the Navier–Stokes equation with random forces, so that stochastic partial differential equations then come into play. There is a huge literature in this subject, after the pioneering mathematical work [6]. Stochasticity is typically introduced at a Eulerian level, although some stochastic Lagrangian models of Langevin type (with smooth Lagrangian trajectories and stochastic velocities) have also been considered in turbulence ([7]).

More recently stochastic advection by Lie transport was introduced by D.D. Holm in [8]. The resulting equations of motion are stochastic partial differential equations and the approach is also Eulerian.

It would be impossible to mention here all stochastic approaches to fluid dynamics. We have only chosen some topics and a few corresponding references. It is worthwhile to mention that there are some interesting probabilistic representation formulae. Representing solutions of partial differential equations as expected values of functionals of stochastic processes is a tradition in the field of stochastic analysis and has also been considered for fluid dynamics. Very roughly speaking, we can find three different approaches: the probabilistic representation of the vorticity field as in [9], the analysis through branching processes and the Fourier transform as in [10] and that using Lagrangian diffusion processes as in [11].

On the other hand, in this paper, we are concerned with variational principles for deterministic dissipative fluid dynamics, and a brief state-of-the-art review follows. There are relatively few references on the special stochastic view of deterministic fluid dynamics advocated here. The oldest we know are [12,13], where the Laplacian term of the Navier–Stokes equation was interpreted, quite informally, as the presence of an underlying Brownian motion. In the famous paper [14], as well as in [15,16], a rigorous geometric strategy for the Euler equation was developed. These geometric ideas have been extended by us to the Navier–Stokes equation, together with the associated variational principle, giving rise to a new stochastic geometric approach to dissipative dynamics. Recently, [17,18], more physical-oriented works, were influenced by references [12,13] and also by our work.

The review paper presented here is, of course, very selective, not in terms of the qualities of quoted references, but in terms of their perspectives, in relation with the interplay between deterministic and stochastic viewpoints. We apologize to the many authors whose important works we were not able to describe here. We hope, however, to describe a special viewpoint in a consistent manner.

In a nutshell, the goal of this review is to show that probabilistic methods play a central role in the study of the deterministic Navier–Stokes equation, both from a conceptual and a practical point of view. Indeed, on the one hand the solutions to that equation satisfy stochastic variational principles. On the other hand, since the solutions to the Navier–Stokes equation may be interpreted as drifts of diffusion processes, one may import many techniques from stochastic analysis to investigate them. Those techniques include the use of stochastic differential equations, of forward–backward stochastic systems and of related numerical techniques.

2. Results

As a dissipative system, there are well known obstructions if we want to derive the classic Navier–Stokes equation from a deterministic variational principle. Allowing the Lagrangian paths to be random and using Itô differential calculus (c.f [19]), we show how we can still derive the Navier–Stokes equation, without any randomness in the external forces, from a (stochastic) variational principle, where the Lagrangian functional is still the classic one, but computed over random Lagrangian trajectories. In some sense, the equation can be regarded as a generalized geodesic equation defined in some suitable space and the variational principle reduces to Hamilton's principle for the incompressible Euler equation when the viscosity vanishes. Our variational approach can be extended in order to consider stochastic Navier–Stokes equations as well.

After recalling Arnold's variational approach to the Euler equation, we describe in Section 4 two stochastic variational principles for the Navier–Stokes equation. The first is a "direct" generalization of Arnold's variational principle to the viscous case, the viscosity being associated with the random behavior of the particles. Incompressibility is incorporated in the definition of their trajectories. The second imposes incompressibility via a Lagrange multiplier.

Having justified our approach of Navier–Stokes equation using stochastic Lagrangian paths, Section 5 is devoted to several possible mathematical methods to study such paths. We can use the theory of forward–backward stochastic differential equations: this is explained in Section 5.1. We can also use entropy methods, since our action functional is essentially given by an entropy quantity. A notion of weak solutions, in the spirit of Brenier's work for the Euler equation and of optimal transport theory, allows us to consider cases where other methods are not accessible by lack of regularity. In Section 6 we describe some stability properties of the Navier–Stokes stochastic Lagrangian flows. The following paragraph is devoted to stochastic perturbations of the Navier–Stokes equation: we show that they can also be derived from a variational principle. Other equations and methods are mentioned in Section 8, as well as some future research problems. Finally a brief appendix contains basic notions of Itô stochastic calculus that are used in this paper.

3. The Non Viscous Case

As pointed out by Arnold ([14]), the incompressible Euler equation, corresponding to the case where the viscosity of the fluid is equal to zero, is an equation of geodesics on a suitable (infinite dimensional) space of functions. If we consider Lagrangian paths $g(t,x)$ such that $\frac{\partial}{\partial t}g(t,x) = u(t,g(t,x))$, with u solution of the Euler equation

$$\frac{\partial}{\partial t}u + (u.\nabla)u = -\nabla p, \quad \text{div}u = 0, \tag{2}$$

the acceleration $\frac{\partial^2}{\partial t}g$ is equal to a gradient function and thus, at every time, orthogonal (for the L^2 scalar product) to vector fields of zero divergence. This means that, if we endow the space of diffeomorphisms preserving the volume measure m of the underlying configuration space a structure of manifold, the Lagrangian flows $g(t,\cdot)$ will be geodesics in such a manifold since its tangent space will consist of divergence free vector fields. In particular they will be critical points of the action functional defined by the kinetical energy:

$$S[g] = \frac{1}{2}\int_0^T \int |\dot{g}(t,x)|^2 dm(x)dt = \frac{1}{2}\int_0^T \|\dot{g}(t)\|^2_{L^2(dm)}dt. \tag{3}$$

More precisely, Arnold showed that the Euler equation above corresponds to the equation of the geodesic flow of the (right-invariant) L^2 metric on the group of diffeomorphisms of the underlying configuration space that preserve the volume measure and have a certain Sobolev regularity. The velocity can be recovered from the Lagrangian flow: $u(t,x) = (\frac{\partial}{\partial t}g_t)(g_t^{-1}(x))$.

This program was rigorously developed in [15] and geodesics were shown to exist locally and under certain regularity restrictions on the initial conditions. Moreover the information on the geometry of the problem had important consequences, for instance in describing the chaotic behavior of Euler Equation (c.f [16]) and its consequences for weather prediction. Instability of the geodesic Euler flows can be described in terms of the sectional curvatures of the group of volume preserving diffeomorphisms. Explicit estimates for the curvatures, being non positive, show that, essentially, the weather is unpredictable.

Notice that this is a special case of Lagrangian system treated in Geometric Mechanics via variational principles on Lie groups ([20]). Indeed the space of volume preserving diffeomorphisms has also a group structure for the composition of maps and the L^2 metric is right-invariant.

4. The Deterministic Navier–Stokes Equation

The situation concerning variational principles for the Navier–Stokes equation is radically different since, contrary to the Euler one, this system is dissipative. Our approach describes dissipation through an introduction of noise in the Lagrangian trajectories. We replace deterministic Lagrangian paths by semimartingales of the form

$$d\xi_t(x) = dM_t(x) + D_t\xi_t(x), \qquad \xi_0(x) = x \tag{4}$$

where M_t is a martingale (for instance, a Brownian motion) and $D_t\xi_t$ denotes the bounded variation part of the semimartingale. The use of the notation D_t is not an accident, since it corresponds to a mean derivative in time (recall that, almost-everywhere, ξ_t satisfying Equation (3) will not be time-differentiable). Consider the definition of the generalized derivative D_t. Namely, for every regular function F,

$$D_tF(t,\xi_t) = \lim_{\epsilon\to 0}\frac{1}{\epsilon}E_t[F(t+\epsilon,\xi_{t+\epsilon}) - F(t,\xi_t)], \tag{5}$$

where E_t denotes the conditional expectation with respect to the past information of the process ξ. Applying this definition to the identity function, and since the operator D_t vanishes when considered on martingales, we see that $D_t\xi_t$ coincides with the drift of the process and can indeed be understood as a regularized derivative. This notion has been known in Stochastic Analysis since its beginnings, as it corresponds to the definition of the generator of the process ξ_t, but it became more relevant in dynamics with the works of Nelson [21].

When ξ takes values in a non-Euclidean space, notably a Riemannian manifold or a Lie group, one can extend the definition of generalized derivative using parallel transport. In this paper we will mainly consider the Euclidean setting. We denote the configuration space by $\mathcal{O}$ and, for now, we assume that $\mathcal{O}$ has no boundary (case with periodic boundary conditions, for example).

4.1. *Stochastic Geometric Mechanics Approach*

We consider the space G_V of bijective maps $g : \mathcal{O} \to \mathcal{O}$ which belong to $L^2 = L^2(dm)$ and keep the volume measure invariant, namely such that $\int f(g(x))dm(x) = \int f dm(x)$ for every (continuous) function f. This infinite dimensional space of maps has a group structure under composition and a Riemannian one with respect to the L^2 metric ([15]). The tangent space to the identity map consists of vector fields on $\mathcal{O}$ with zero divergence. Moreover a simple covariant derivative can be defined: for such two vector fields X and Y, $\nabla_X Y = \Pi(\partial_X Y)$ where ∂ is the usual Lie derivative and Π the projection operator from L^2 to divergence free vectors associated to the Helmholtz (or Hodge, according to the source) decomposition.

The next result was proved in [22], then extended to Riemannian manifolds in [23] and formulated on general Lie groups in [24].

We consider a stochastic action functional where the Lagrangian is the kinetic energy computed on the generalized derivative of a semimartingale. The corresponding norm is the L^2 one. More precisely, for a G_V-valued semimartingale ξ as in Equation (4) we define

$$S[\xi] = \frac{1}{2} E \int_0^T \int_{\mathcal{O}} |D_t \xi_t(x)|^2 dt dm(x) \tag{6}$$

where E means expectation with respect to the underlying probability. A particular class of such semimartingales is the following. To a time dependent vector field with zero divergence $u(t,\cdot)$, $t \in [0,T]$, and belonging to L^2, we associate the stochastic differential equation (c.f [19], for example, as a reference for Itô stochastic calculus, as well as the Appendix A in this article),

$$dg_t^u(x) = \sqrt{2\nu}\, dW_t + u(t, g_t^u(x))dt, \qquad g_0^u(x) = x \tag{7}$$

where W is a standard Brownian motion. This equation defines a stochastic flow of maps on $\mathcal{O}$ that belong to G_V and satisfy $D_t g_t^u = u(t, g_t)$.

Considering a simple Brownian motion may be regarded as oversimplifying. Actually one should introduce martingales driven by vector fields modeling the correlations observed in the physical model ([8]). Typically Lagrangian paths are of the form

$$dg_t = \sum_k H_k(g_t) dW_t^k + u(t, g_t)dt$$

where H_k are correlation eigenvectors and W^k independent Brownian motions. Our approach covers such cases ([22,24,25]) and we chose here a Brownian motion only to simplify the exposition.

We are interested in derivating $S[g^u]$; in particular we want to consider variations of the paths g^u for which the functional above is still well defined, i.e., they are still G_V-valued semimartingales. Consider the exponential type functions

$$e_t(\epsilon v)(x) = x + \epsilon \int_0^t \dot{v}(s, e_s(\epsilon v)(x))ds$$

with $\epsilon > 0$ and where $v(t,\cdot)$ is a smooth time dependent vector field such that $v(0) = v(T) = 0$ and div $v(t,\cdot) = 0$ for every $t \in [0,T]$. Notice that, up to the first order in ϵ, we have $e_t(\epsilon v)(x) \simeq x + \epsilon v(t,x)$. The variations of the paths $g^u(t)$ will be defined by left composition, since the functional is right-invariant:

$$g_t^{u,\epsilon} = e_t(\epsilon v) \circ g^u(t)$$

We have, using Itô calculus,

$$dg_t^{u,\epsilon} = \nabla e_t(\epsilon v)(g_t^u)\sqrt{2\nu} dW_t + [\dot{e}_t(\epsilon v) + (u.\nabla)e_t(\epsilon v) + \nu \Delta e_t(\epsilon v)](g_t^u)]dt$$

By the definition of $e_t(v)$,

$$\left.\frac{d}{d\epsilon}\right|_{\epsilon=0} S[e.(\epsilon v) \circ g^u(\cdot)] = E \int_0^T \left(\int D_t g^u(t)(x).D_t v(g^u(t)(x))dm(x) \right) dt$$

and by Itô's formula,

$$d \int D_t g^u(t)(x).v(g^u(t)(x))dm(x) = \int dDg^u(t)(x).v(g^u u(t)(x))dm(x) + \int Dg^u(t)(x).dv(g^u(t)(x))dm(x)$$

$$+ \int dD_t g^u(t)(x).dv(g^u(t)(x))dm(x)$$

The last (Itô's contraction) term is equal to

$$2\nu(\int(\nabla v\otimes\nabla u)(g^u(t)(x))dm(x))dt$$

As $v(0)=v(T)=0$ this implies,

$$\frac{d}{d\epsilon}\Big|_{\epsilon=0} S[e.(\epsilon v)\circ g^u(\cdot)]=-E\int_0^T(\int(D_tD_tg^u(t)(x)dm(x))dt-2\nu E\int_0^T(\int(\nabla v\otimes\nabla u)(g^u(t)(x))dm(x))dt$$

On the other hand

$$D_tD_tg^u(t)=D_tu(t,g^u(t))=(\frac{\partial}{\partial t}u+(u.\nabla)u+\nu\Delta u)(g^u(t)).$$

Therefore, using the invariance of the measure with respect to the process g^u and integration by parts, we obtain

$$\frac{d}{d\epsilon}\Big|_{\epsilon=0} S[e.(\epsilon v)\circ g^u(\cdot)]=-E\int_0^T\int\left(([\frac{\partial}{\partial t}u+(u.\nabla)u-\nu\Delta u].v)(t,g^u(t)(x))dm(x)\right)dt$$

$$=-\int_0^T\left(\int([\frac{\partial}{\partial t}u+(u.\nabla)u-\nu\Delta u].v)(t,x)dm(x)\right)dt$$

for every v with zero divergence, which means that $\frac{\partial}{\partial t}u+(u.\nabla)u-\nu\Delta u$ is the gradient of some function.

We have therefore the following

Theorem 1. *Let $u(t,\cdot)$ be a smooth time dependent divergence-free vector field defined on $[0,T]\times\mathcal{O}$. Let $g^u(t)$ be a stochastic Brownian flow with diffusion coefficient $\sqrt{2\nu}$ and drift u (as in Equation (7)). Then g^u is a critical point of the energy functional S if and only if there exists a function p such that $u(t)$ verifies the incompressible Navier–Stokes Equation (1).*

Remark 1. *When $\mathcal{O}$ is a Riemannian manifold (say, without boundary), we can still define an action functional of the form in Equation (6), but some more concepts are needed.*

In that case the Itô differential of an $\mathcal{O}$-valued semimartingale Y is defined by

$$dY_t=P(Y)_t\,d\left(\int_0^{\cdot}P(Y)_s^{-1}\circ dY_s\right)_t$$

where $P(Y)_t:T_{Y_0}M\to T_{Y_t}M$ is the parallel transport associated with the Levi–Civita connection along $t\mapsto Y_t$. Alternatively, in local coordinates,

$$dY_t=\left(dY_t^i+\frac{1}{2}\Gamma^i_{jk}(Y_t)dY_t^j\otimes dY_t^k\right)\partial_i$$

where Γ^i_{jk} are the Christoffel symbols of this connection.

If the semimartingale Y_t has an absolutely continuous drift, we denote it by DY_t: for every 1-form $\alpha\in\Gamma(T^*\mathcal{O})$, the finite variation part of $\int_0^{\cdot}\langle\alpha(Y_t),\,dY_t\rangle$ is $\int_0^{\cdot}\langle\alpha(Y_t),\,DY_t\,dt\rangle$. We consider an incompressible Brownian flow $g^u(t)$ with covariance $a\in\Gamma(T\mathcal{O}\otimes T\mathcal{O})$ and time dependent drift $u(t,\cdot)\in\Gamma(T\mathcal{O})$. We assume that for all $x\in\mathcal{O}$, $a(x,x)=2\nu\mathbf{g}^{-1}(x)$ for some $\nu>0$, where $\mathbf{g}$ denotes the metric tensor of the manifold. Such incompressible flows are known to be well defined on compact symmetric spaces and on compact Lie groups.

This means that

$$dg_u(t)(x) \otimes dg_u(t)(y) = a\left(g_u(t)(x), g_u(t)(y)\right) dt,$$
$$dg_u(t)(x) \otimes dg_u(t)(x) = 2\nu \mathbf{g}^{-1}\left(g_u(t)(x)\right) dt,$$

the drift of $g^u(t)(x)$ is absolutely continuous and satisfies $Dg^u(t)(x) = u(t, g_u(t)(x))$. Then, using the same kind of variations as in the flat case, we derive (c.f [23]), from the energy functional

$$S(g^u) = \frac{1}{2}E\left[\int_0^T \left(\int_{\mathcal{O}} |Dg_u(t)(x)|^2\, dm(x)\right) dt\right],$$

the equation

$$\frac{\partial}{\partial t}u + \nabla_u u = \nu L u - \nabla p$$

where $L = dd^* + d^*d$ is the the Laplace–de Rham operator. We recall that when computed on forms and, in particular, on vector fields, L differs from the usual Levi–Civita Laplacian by a Ricci curvature term.

4.2. Lagrange Multipliers Approach

Here we describe another stochastic variational principle for the Navier–Stokes equation, that uses a Lagrange multiplier formulation. Its advantage is that it does not need to be formulated in the space G_V, since incompressibility is not incorporated in the definition of the flows, but given instead by the multiplier condition. The problem becomes strictly finite-dimensional and variations of the paths may be defined simply by shifts. It also allows us to consider domains with boundary. References for the Lagrange multipliers approach are [26,27], where one can find detailed proofs. The first concerns the case of a domain without boundary (more precisely the torus) and in the second one can find its extension to a domain with boundary.

Let $\mathcal{O}$ be a domain with a regular boundary. We consider the Navier–Stokes equation with Neumann boundary condition on $[0, T] \times \mathcal{O}$, namely the condition $\nabla u.n = 0$, where n denotes the unit vector normal to the boundary. The stochastic Lagrangian flows can now take into account reflections at the boundary of the domain and can be written,

$$dg_t(x) = \sqrt{2\nu}dW_t + u(t, g_t(x))dt + n(g_t(x))d\ell(t), \qquad g_0(x) = x \in \bar{\mathcal{O}} \tag{8}$$

where $\ell(t) = \int_0^t 1_{\delta\mathcal{O}}(g_s(x))d\ell(s)$ is the local time, representing the amount of time spent by the diffusion process in the neighborhood of points in the boundary.

The action functional is defined as

$$S(g, p) = \frac{1}{2}E\int_0^T \int |D_t g_t(x)|^2 dtdm(x) + E\int_0^T \int p(t, g_t(x))(\det \nabla g_t(x) - 1)dtdm(x) \tag{9}$$
$$:= S^1(g, p) + S^2(g, p) \tag{10}$$

The extra term S^2 corresponds to a Lagrange multiplier whose constraint forces the paths to keep the volume measure preserved during the evolution (incompressibility condition). The variable p is defined in the linear space $L^2([0, T] \times \mathcal{O})$. We consider variations of the form

$$g_t(\cdot) \to g_t^\epsilon(\cdot) = g_t(\cdot) + \epsilon h(t, g_t(\cdot))$$

$$p(t, \cdot) \to p^\epsilon(t, \cdot) = p(t, \cdot) + \epsilon\varphi(t, g_t(\cdot))$$

with $h(t, x)$ and $\varphi(t, x)$ deterministic and smooth, satisfying $h(T, \cdot) = h(0, \cdot) = 0$, $h = 0$ on $\partial\mathcal{O}$.

Concerning S^2, we have,

$$\left.\frac{d}{d\varepsilon}\right|_{\varepsilon=0} S^2(g^\epsilon,p^\epsilon) = E\int_0^T\int \varphi(t,g_t(x))(\det\nabla g_t(x)-1)dtdm(x) \tag{11}$$

$$+E\int_0^T\int(\nabla p(t,g_t(x)).h(t,g_t(x))(\det\nabla g_t(x)-1)dtdm(x) \tag{12}$$

$$+E\int_0^T\int p(t,g_t(x))\left.\frac{d}{d\varepsilon}\right|_{\varepsilon=0}\det\nabla(g_t(x)+\epsilon h(t,g_t(x))dtdm(x) \tag{13}$$

Since φ is arbitrary we conclude from the first term of Equation (11) that critical points of the action are volume-preserving diffeomorphisms ($\det\nabla g_t(x)=1$) and therefore have divergence-free drifts. It follows immediately that Equation (12) is also equal to zero. The computation of the third term gives

$$\left.\frac{d}{d\varepsilon}\right|_{\varepsilon=0} S^2(g^\epsilon,p^\epsilon) = -E\int_0^T\int(\nabla p(t,g_t(x)).\,h(t,g_t(x)))dtdm(x).$$

Actually this last computation does not use stochastic calculus:

$$\left.\frac{d}{d\varepsilon}\right|_{\varepsilon=0}\det\nabla(g_t(x)+\epsilon h(t,g_t(x)))=\det\nabla g_t(x)\,\mathrm{tr}\Big((\nabla g_t(x))^{-1}\left.\frac{d}{d\varepsilon}\right|_{\varepsilon=0}\nabla g_t^\epsilon(x)\Big)$$

$$=\det(\nabla g_t(x))\,\mathrm{tr}\big((\nabla g_t(x))^{-1}\nabla(h(t,g_t(x)))\big).$$

Since

$$\partial_i(p(t,g_t)(\nabla g_t)^{-1}_{ij}h(t,g_t)^j)=\partial_i(p(t,g_t))(\nabla g_t)^{-1}_{ij}h(t,g_t)^j+p(t,g_t)(\nabla g_t)^{-1}_{ij}\partial_i(h^j(t,g_t))$$
$$+p(t,g_t)h^j(t,g_t)\partial_i(\nabla g_t)^{-1}_{ij},$$

$$(13)=-E\int_0^T\int[\partial_i(p(t,g_t))(\nabla g_t)^{-1}_{ij}+p(t,g_t)\partial_i((\nabla g_t)^{-1}_{ij})]h^j(t,g_t)\det\nabla g_t\,dtdx.$$

Notice that we already concluded that $\det\nabla g_t=1$. On the other hand,

$$\sum_i\partial_i(\nabla g_t)^{-1}_{ij}=0.$$

Indeed, derivating the equality $\det\nabla g_t=1$, we get

$$\partial_k\det(\nabla g_t)=\ \mathrm{tr}\big((\nabla g_t)^{-1}\partial_k(\nabla g_t)\big)=\sum_i(\nabla g_t)^{-1}_{ij}\partial_k\partial_i g_t^j=0.$$

Also, derivating equality $(\nabla g_t)^{-1}_{ij}\partial_k g_t^j=\delta_{ik}$, we obtain

$$\sum_i\partial_i(\nabla g_t)^{-1}_{ij}\partial_k g_t^j+(\nabla g_t)^{-1}_{ij}\partial_i\partial_k g_t^j=0;$$

therefore

$$\sum_i\partial_i(\nabla g_t)^{-1}_{ik}=-((\nabla g_t)^{-1}_{ij}\partial_k\partial_i g_t^j)(\nabla g_t)^{-1}_{jk}=0$$

and

$$(13)=-E\int_0^T\int(\partial_i(p(t,g_t(x)))(\nabla g_t(x))^{-1}_{ij})h_t^j(g_t(x))\det\nabla g_t(x))dtdm(x)$$

$$= -E\int_0^T \int (\nabla p(t, g_t(x)).h(t, g_t(x)))dtdm(x).$$

We now look at the derivation of S^1. We can prove, using Itô calculus and similarly to the computation in Section 4.1, that

$$\frac{d}{d\varepsilon}\bigg|_{\varepsilon=0} S^1(g^\epsilon, p^\epsilon) = E < Dg_T, h(T, g_T) > -E < Dg_0, h(0, g_0) > -E\int_0^T \int (D_t D_t g_t(x).h(t, g_t(x)))dtdx$$

$$-E\int_0^T \int (dD_t g_t(x).dh(t, g_t(x)))dx.$$

$$= -E\int_0^T \int (D_t D_t g_t(x).h(t, g_t(x)))dtdm(x) - 2\nu E\int_0^T (\nabla u.\nabla h)(t, g_t(x))dtdm(x)$$

Using equality $D_t D_t g_t(x) = (\frac{\partial}{\partial t}u + (u.\nabla)u + \nu\Delta u + n\nabla u\ \ell(t))(t, g_t(x))$ and integration by parts, we deduce that

$$\frac{d}{d\varepsilon}\bigg|_{\varepsilon=0} S^1(g^\epsilon, p^\epsilon) = -E\int_0^T \int [((\frac{\partial}{\partial t}u + (u.\nabla)u - \nu\Delta u).\ h) + (n.\nabla u)\ h\ \ell(t)](t, g_t(x))dtdm(x).$$

Combining the expressions above for the variation of the action functional and using the invariance of the volume measure for the flows, we obtain the following result,

Theorem 2. *A diffusion g_t of the form of Equation (8) and a function p are critical for the action functional in Equation (9) if and only if the drift $u(t,\cdot)$ of g_t satisfies the Navier–Stokes equation*

$$\partial_t u + (u.\nabla)u = \nu\Delta u - \nabla p, \qquad div\, u(t,\cdot) = 0, \quad \nabla u.n = 0 \text{ in } \partial\mathcal{O} \tag{14}$$

with $t \in [0, T]$.

5. Constructing Solutions of Navier–Stokes Equation by Probabilistic Methods

5.1. Forward–Backward Stochastic Differential Systems

The stochastic Lagrangian flows $g_t(\cdot)$ obtained either via the geometric or the Lagrangian multiplier approach described above satisfy an equation of Newton type. More precisely, on a fixed time interval $[0, T]$ and after the change of time $v(t, x) = -u(T - t, x)$, we have $Dg_t(x) = -v(T - t, g_t(x))$ and

$$D_t v(t, g_t(x)) = -\nabla p(t, g_t(x)), \qquad g_0(x) = x \tag{15}$$

or, regarded as an equation in the weak L^2 sense,

$$D_t v(t, g_t(\cdot)) = 0, \qquad g_0 = \text{id}. \tag{16}$$

Formally, when the viscosity is zero, this is Arnold's geodesic equation and ours can be seen, indeed, as a generalized geodesic equation. Moreover, although we only consider in this paper an Euclidean setting, the framework can be extended without essential difficulties to a general Riemannian manifold or a Lie group ([23,24]).

How does one solve directly this equation of motion? One possible way is to characterize it in terms of forward–backward stochastic differential systems, which are second order stochastic equations (c.f the Appendix A). We have formulated the problem in [28] first, then solved it in [29,30] for some specific function spaces and in two and three dimensions, respectively. The characterization in terms of forward–backward stochastic differential equations has also the advantage that it may allow to implement numerical methods, known for such systems (c.f for example, [31] and the recent work [32]).

The forward–backward system solved by our stochastic Lagrangian flows can be written in the form

$$\begin{cases} dg_t = \sqrt{2\nu}\, dW_t + Y_t dt \\ dY_t = Z_t dW_t - \nabla p(g_t) dt \end{cases}$$

together with a given initial condition for the forward equation ($g_0 = x$) and a final condition for the backward one, Y_T. It can be proved these type of systems are well posed and their solutions are of the form $Y_t = Y_t(x) = v(t, g_t(x))(= -u(t, g_t(x)))$ for some vector field v. We have,

$$D_t Y_t = D_t D_t g_t = -\nabla p(g_t)$$

(compare with Equation (15)). Variable Z, although a priori unknown, is a posteriori determined and equal to $\sqrt{2\nu}\nabla v(g_t)$.

To be more precise, let u be a solution of the Navier–Stokes equation in the time interval $t \in [0, T]$ and assume that u is regular. Let $g_s^t(x)$ be the unique solution of the following stochastic differential equation,

$$\begin{cases} dg_s^t(x) = \sqrt{2\nu} dW_s - u(T-s, g_s^t(x)) ds \\ g_t^t(x) = x, \end{cases}$$

with $s > t$. We define $Y_s^t(x) = u(T-s, g_s^t(x))$, $Z_s^t(x) = \nabla u(T-s, g_s^t(x))$; applying Itô's formula directly, the following forward–backward stochastic differential system with solution $(g_s^t(x), Y_s^t(x), Z_s^t(x), u(t,x), p(t,x))$ is derived,

$$\begin{cases} dg_s^t(x) = \sqrt{2\nu} dW_s - u(T-s, g_s^t(x)) ds \\ dY_s^t(x) = \sqrt{2\nu} Z_s^t(x) dW_s - \nabla p(T-s, g_s^t(x)) ds \\ Y_t^t(x) = u(T-t, x) \\ g_t^t(x) = x,\ Y_T^t(x) = u_0(g_T^t(x)) \end{cases} \tag{17}$$

Recall also that the pressure satisfies the following identity

$$\Delta p(t,x) = \sum_{i,j=1}^{3} \partial_i u^j(t,x) \partial_j u^i(t,x). \tag{18}$$

On the other hand, if $(g_s^t(x), Y_s^t(x), Z_s^t(x), u(t,x), p(t,x))$ is a solution of Equation (17) together with Equation (18) and u is regular enough, then the vector field $u(t,x) := Y_{T-t}^{T-t}(x)$ satisfies the Navier–Stokes equation for $t \in [0, T]$. In particular, we can show that div $u(t,x) = 0$ due to the expression of $p(t,x)$ given by Equation (18).

In order to incorporate Equation (18) in the forward–backward system and obtain a closed system of equations we proceed as follows. Denote by N the Newton's potential in $\mathbb{R}^d, d \geq 3$, i.e., the operator Δ^{-1} which is given by

$$Nf(x) = C(d) \int_{\mathbb{R}^d} \frac{f(y)}{|x-y|^{d-2}} dy,$$

where $C(d)$ is a constant depending on the dimension d. Then we can write Equations (17) and (18) as

$$\begin{cases} dg_s^t(x) = \sqrt{2\nu} dW_s - u(T-s, g_s^t(x)) ds \\ dY_s^t(x) = \sqrt{2\nu} Z_s^t(x) dW_s - \nabla N\Big(\sum_{i,j} \partial_i v^j - \partial_j v^i\Big)(T-s, g_s^t(x)) ds \\ Y_t^t(x) = u(T-t, x) \\ g_t^t(x) = x, Y_T^t(x) = u_0(g_T^t(x)), \end{cases} \tag{19}$$

Using suitable L^p bounds of the operator $\nabla N\Big(\sum_{i,j}\partial_i v^j - \partial_j v^i\Big)$ we have constructed in [30] local unique solutions of the system in Equation (19), and therefore of the Navier–Stokes equation, in some Sobolev-type functional spaces in dimension $d \geq 3$. The two-dimensional case was studied in a similar way [29], but via the vorticity equation.

Such type of forward–backward differential equations were also studied on general Lie groups in [33].

5.2. Entropy Methods

In a closely related recent work ([34]), we explored the fact that the stochastic kinetic energy used here coincides in fact with a relative entropy, of the law of the diffusion process associated with Navier–Stokes in relation to the Wiener law. Brenier's relaxation of Arnold's geodesic problem, extended to the stochastic setting, becomes, in this sense, an entropy minimization problem.

Recall that the relative entropy of a probability measure Q on some measurable set X with respect to a reference probability measure R on X is defined as

$$H(Q|R) = \int_X \log\Big(\frac{dQ}{dR}\Big) dQ$$

when Q is absolutely continuous with respect to R. What we have named the Brenier–Schrödinger problem in [34] consists of the following. Consider X to be an Euclidean space and fix as reference measure R as the law of the reversible Brownian motion on a time interval $[0,T]$, namely $R = \int_X P^x dm(x)$ for P^x the law of a Brownian motion starting at x. The we look for a measure Q minimizing the relative entropy $H(Q|R)$ under the constraint that Q_t, the marginals at time t, are given, and that the joint law at initial and final times, Q_{0T}, is also given. In our case the relevant condition for the marginals is the incompressibility condition $Q_t = m$ for all times.

This is a convex minimization problem reminiscent of the Schrödinger problem in Quantum Mechanics ([35–37]) and, simultaneously, a generalization of a Monge–Kantarovich mass transportation problem (c.f [38]).

The relation between the entropy problem and our stochastic variational one relies on the fact that, by Girsanov's theorem (c.f for instance [19]), if the Lagrangian path is given by the stochastic differential equation $dg_t^u = \sqrt{2\nu}dW_t + u(t, g_t^u)dt$ with initial condition R, and if Q denotes the law of this stochastic process, we have

$$H(Q|R) = H(Q_0|R) + \frac{1}{2}E\int_0^T |u(t, g_t^u)|^2 dt$$

Therefore the entropy minimization problem corresponds to our stochastic variational principle, described here in Section 4. We can use then tools of convex analysis, establish the existence of a solution, consider a dual problem that provides the pressure term and exhibit Lagrange multipliers. Since our solution is the law of a stochastic process which is absolutely continuous with respect to the reference measure, one can write its Radon–Nikodym derivative and compare it to an alternative expression obtained from convex duality. We find that the drifts of our Lagrangian flows, written as velocity fields, have unusual properties: when stratified (conditioned to the arrival point of the stochastic process), the backward velocity is the gradient of the solution to a Hamilton–Jacobi–Bellman equation, while the pressure remains independent of the arrival point. We recover, with this method, a structure that is analogous to Brenier's stratified solutions in [39].

5.3. A Weak Notion of Navier–Stokes Solutions

One cannot expect that the stochastic variational approach will determine solutions of the Navier–Stokes in full generality. It is already the case that geodesics associated with Euler equation do not exist in some situations. The main difficulty is that the topology induced by the energy is not strong enough to deal with the needed regularity of the Lagrangian maps. In order to overcome the problem, Brenier has introduced a concept that he named "generalized solutions", replacing the notion of geodesic path by a probability measure over geodesic paths, in the spirit of the Monge–Kantarovich problem ([40]). An extension of Brenier's framework to the stochastic Lagrangian setting (and in particular to the Navier–Stokes equation) can be found in [41].

For the Euler equation, Brenier looked for probability measures P on the path space $C([0,T];\mathcal{O})$ which minimize the energy functional

$$\frac{1}{2}\int_{C([0,T];\mathcal{O})}\Big[\int_0^T |\dot{\gamma}(t)|^2 dt\Big] dP(\gamma)$$

with the incompressibility constraints $(e_t)_* P = dm(x)$, where $e_t : \gamma \to \gamma(t)$ is the evaluation map. A solution of such problem gives rise to a weaker type of Lagrangian flows for the Euler equation. Indeed one can define a probability measure μ on $[0,T] \times \mathcal{O}^2$ by

$$\int f(t,x,v) d\mu = \frac{1}{T}\int_{C([0,T];\mathcal{O})} f(t,\gamma(t),\gamma'(t)) dP(\gamma) dt$$

and μ solves the Euler equation in a weak sense, namely

$$\int [v.w'(x)\alpha'(t) + v.(\nabla w(x).v)\alpha(t)] d\mu(t,x,v) = 0$$

holding for all smooth test functions $\alpha(t)$ and every smooth divergence free vector field v. These kind of weak solutions of partial differential equations is known as solutions in the sense of Di Perna and Majda.

In the viscous case, we consider $\mathcal{O}$-valued semimartingales g_t of the form $dg_t = \sqrt{2\nu} dW_t + u_t dt$ (remark that the drift can be random) and their corresponding laws $\mathbb{P}^g$ on the path space $C([0,T];\mathcal{O})$: for every cylindrical functional F,

$$\int_{C([0,T],\mathcal{O})} F(\gamma(t_1),...,\gamma(t_n)) d\mathbb{P}^g(\gamma) = \int_{\mathcal{O}}\Big[\int_{C([0,T],\mathcal{O})} F(g_{t_1}(x),...,g_{t_n}(x)) d\mathbb{P}^g_x\Big] dm(x)$$

where $\mathbb{P}^g = \mathbb{P}^g_x \otimes dm(x)$. Under $\mathbb{P}^g_x$, the semimartingale g_t starts from x.

We say that the semimartingale g_t is incompressible if, for each $t > 0$,

$$E_{\mathbb{P}^g}[f(g_t)] = \int_{\mathcal{O}} f(x) dx, \quad \text{for all } f \in C(\mathcal{O})$$

and we define the energy functional of a semimartingale g by the formula

$$\frac{1}{2}\int_{\mathcal{O}} E_{\mathbb{P}^g_x}\Big(\int_0^T |D_t g(x)|^2 dt\Big) dm(x) = \frac{1}{2}\int_{\mathcal{O}} E_{\mathbb{P}^g_x}\Big(\int_0^T |u_t(g_t(x))|^2 dt\Big) dm(x)$$

Considering suitably admissible variations we concluded in [41] that the semimartingales which are critical points of the energy functional above solve the Navier–Stokes equation in the following weak form:

$$\int_{\mathcal{O}}\int_0^T < u_t, \alpha'(t) w + \alpha(t)\, \nabla w \cdot u_t - \nu\, \alpha(t) \Delta w > \; dt dm(x) = 0 \tag{20}$$

for all smooth functions of time α and all smooth vector fields w such that $\mathrm{div}(w) = 0$. Moreover we have showed that, in the case where $\mathcal{O}$ is the torus (corresponding to periodic boundary conditions) and under certain additional assumptions, classical solutions of the Navier–Stokes are minimizers of the energy action functional.

6. Stability Properties

In finite dimensions it is well known that the behavior of geodesics can be expressed in terms of the curvature of the underlying manifold via the Jacobi equation.

Arnold's approach to the Euler equation allowed to show, in many cases, that the curvature of the spaces of diffeomorphisms is negative and therefore that the fluid trajectories are unstable (or "chaotic"), i.e., their distance, starting from different initial conditions, grows exponentially during time evolution (c.f [16]).

For viscous flows it is expected that particles become closer and closer after some possible initial stretching. For our model, at least in the case of the two dimensional torus, we could show ([23] and also [42]) that sensitivity with respect to initial conditions of the trajectories is enhanced by their stochasticity. The behavior will depend of the choice of diffusion coefficients that we consider in the Lagrangian paths or, in other words, on which scales and with what strength the motion is excited.

We have considered on the two-dimensional flat torus $\mathbb{T} = (\mathbb{R}/2\pi\mathbb{Z})^2$ a Brownian motion of the form

$$dB_t = \sum_{k \in \mathbb{Z}^2} (k_2, -k_1)\sqrt{\nu}\lambda_k A_k dW_t^k$$

where $A_{2k}(x) = \cos(2k.x)$, $A_{2k+1}(x) = \sin((2k+1).x)$, $k \in \mathbb{Z}^2, x \in \mathbb{T}$. The coefficients λ_k have to be such that the process B is well defined, which is achieved by assuming a certain decay at infinity of those coefficients (see [22]). The Lagrangian stochastic processes associated to the Navier–Stokes equation will be defined as in Equation (7), with W_t replaced by B_t.

Consider the following distance between trajectories:

$$\rho^2(g, \tilde{g}) = \int_{\mathbb{T}} |g(x) - \tilde{g}(x)|^2 dm(x).$$

By using Itô calculus we have deduced, in particular, that under the assumption that the Navier–Stokes solution u satisfies $\nabla u(t,x) \leq c_1 e^{-c_2 t}$, we have an estimate for the distance between trajectories of the form

$$\rho_t \geq \rho_s \exp\left(Z_t + c_3 t - \frac{c_1}{c_2}(1 - e^{-c_2 t})\right)$$

for $s \leq t$, where Z is a 1-dimensional Brownian motion, c_3 another constant which depends on the coefficients λ.

The assumption on the gradient of u implies that the velocity decays to zero at exponential rate. On the contrary the stochastic Lagrangian flows, describing the position of the fluid, get apart exponentially, at least for short times. Moreover, by the explicit expression of the constant c_3 ([23]), we observe that the stochastic Lagrangian trajectories for a fluid with a given viscosity constant tend to get apart faster when the higher Fourier modes (and therefore the smaller length scales) are randomly excited.

We can also show how the rotation of two particles, when their distance is small, becomes more and more irregular as time evolves, with explicit formulae in the torus case.

7. A Stochastic Navier–Stokes Equation

When we consider random forces, Navier–Stokes becomes a stochastic partial differential equation. In this paragraph we formulate some of these type of equations by a variational principle, where the action functional is suitably perturbed.

We define the action functional, which is now a random variable (in particular we remove the expectation) with some extra stochastic terms in the Lagrangian, as

$$S_\omega(g,p) = \frac{1}{2}\int_0^T \int |D_t g_t(x)|^2 dt dm(x) + \int_0^T \int p(t, g_t(x))(\det \nabla g_t(x) - 1) dt dm(x) \quad (21)$$

$$+ \int_0^T \int D_t g_t(x) dM_t^g(x) dm(x) - \sqrt{2\nu}\int_0^T \int D_t g_t(x) dW_t dm(x) \quad (22)$$

where M^g denotes the martingale part of the diffusion process g. We use the same variations as in Section 4.2. The variation of the two extra terms gives

$$\int_0^T \int [(h(t,g_t).\sqrt{2\nu} dW_t) + (D_t g_t.(\nabla h(t,g_t).dW_t)) - (h(t,g_t).\sqrt{2\nu} dW_t)] dm(x)$$

that reduces to

$$\int_0^T \int v(t, g_t(x)).(\nabla h(t, g_t(x)).dW_t) dx = \int_0^T v(t,x).(\nabla h(t,x).dW_t)$$

$$= -\int_0^T ((\nabla v(t,x).h(t,x)).dW_t)$$

equality which holds for all h, P-almost surely.

Hence we obtain the following (c.f [25,26]),

Theorem 3. *A diffusion g_t of Equation (8) and a function p are critical for the action functional S_ω if and only if the drift $u(t,\cdot)$ of g_t satisfies the stochastic Navier–Stokes equation*

$$d_t u + (u.\nabla) u dt = \sqrt{2\nu} \nabla u dW_t + \nu \Delta u dt - \nabla p dt, \qquad \text{div}\, u(t,\cdot) = 0, \quad \nabla u.n = 0 \; in \; \partial\mathcal{O} \quad (23)$$

with $t \in [0,T]$.

The stochastic Navier–Stokes equation written above can also be regarded as a stochastic perturbation of Euler equation in the sense that, replacing Itô differentials by Stratonovich ones, Equation (23) is equivalent to

$$d_t u + (u.\nabla) u dt = \sqrt{2\nu} \nabla u \circ dW_t - \nabla p dt, \qquad \text{div}\, u(t,\cdot) = 0, \quad \nabla u.n = 0 \text{ in } \partial\mathcal{O}$$

with $t \in [0,T]$ and where $\circ d$ stands for Stratonovich differential.

We remark that much more general noises can be considered and that we have only chosen a simple Brownian motion for simplicity. On the other hand the fact that we obtain a transport type noise is intrinsically related to our model.

There are many references for stochastic partial differential equations which are perturbations of Euler or Navier–Stokes. One can find in the recent work [5] a model where, as in our case, the noise is multiplicative.

In a general Lie group framework a variational approach to stochastic partial differential equations was developed in [25].

8. Other Equations, Methods and Discussion

The stochastic approach to Navier–Stokes described in this review can, in principle, be applied to describe any second order perturbation of a conservative first-order ordinary differential equation, when the latter is formulated in a suitable geometric way, as the Euler equation is formulated via a geodesic equation. This is the case, for example, for the Camassa–Holm equation, the Hunter–Saxton equation, the average Euler equation and the equations governing the motion of rigid bodies.

The viscous Camassa–Holm equation was studied in detail in [43]. It corresponds to replacing the $L^2(dm)$ metric, the energy in the Lagrangian, by the Sobolev H^1 metric and this equation reads

$$\frac{\partial}{\partial t}v + (u.\nabla)v = \nu\Delta v + \sum_j \nabla u^j.\Delta u^j - \nabla p, \quad \mathrm{div} u = 0$$

where $v = u - \Delta u$.

Compressible Navier–Stokes or viscous Magnetohydrodynamical equations, on the other hand, have to be coupled with tracers (advected quantities) that need to be modeled by extra variables. Mathematically one introduces, together with the Lie group for the Lagrangian flows, a semidirect product vectorial structure (c.f [44]). We refer to [25] for a study of these various equations where we use stochastic methods, and stress again the fact that we can derive such equations by means of variational principles without any reference to thermodynamic considerations.

An extension of Arnold's geometric framework, describing compressible fluid dynamics (and other systems) with Newton's equations on a space of probability densities, can be found in the recent paper [45]. It reveals interesting connections with optimal transport and information theory, in particular.

The construction of solutions as critical points of the stochastic action functionals does not easily follow from the direct methods of the calculus of variations. We have instead indicated an indirect approach based on forward–backward stochastic differential equations, but even there we encountered many technical difficulties. Other possibilities are the use of entropy methods, as described in Section 5.2, or the relaxation of the notion of solution (c.f Section 5.3).

For conservative dynamical systems, it is well known that one can take advantage of the symmetries of such systems to reduce the complexity of the equations. Symmetries also play an important role in the implementation of numerical algorithms designed to investigate the equations in question. The natural objects to be introduced in order to replace constants of motion are martingales, since a martingale M, by definition, is a quantity whose generalized derivative vanishes ($D_t M = 0$). One can find a Noether theorem for a certain class of diffusion processes in [46]. We refer to [47] for a brief discussion on symmetries in our context.

Finally, let us point out that numerical methods tied up with our probabilistic approach still have to be developed.

Funding: This research was partly funded by FCT (Fundação para a Ciência e Tecnologia, Portugal), grant "Schrödinger's problem and Optimal Transport: a multidisciplinary perspective", with reference PTDC/MAT-STA/28812/2017.

Conflicts of Interest: The author declare no conflict of interest.

Appendix A. Some Basic Notions of Stochastic Calculus

We start by recalling the definition of a ($\mathbb{R}$-valued) *Brownian motion* W_t, $t \in [0, T]$. This is a continuous stochastic process (namely, a function of time and of a random parameter in some standard probability space $(\Omega, \mathcal{B}, P)$) that verifies the following properties:

(i) $W_0 = x$;
(ii) W_t has independent increments;
(iii) For $s < t$, $W_t - W_s$ has a normal distribution $\mathcal{N}(0, t - s)$.

A $\mathbb{R}^d$ Brownian motion corresponds to a collection of independent and identically distributed copies of one dimensional Brownian motions. A Brownian motion is also called a Wiener process.

We generally omit the random variable in the notation when we write a stochastic process.

Suppose that the probability space is endowed with a filtration, namely an increasing family $\mathcal{P}_t$ of sub σ-algebras of $\mathcal{B}$. Typically each $\mathcal{P}_t$ represents the events that occur before a time t. A stochastic process M_t is a ($\mathbb{R}$-valued) *martingale* with respect to $\mathcal{P}_t$ if

(i) M_t is $\mathcal{P}_t$-measurable for all t (we also say that M_t is *adapted* to $\mathcal{P}_t$);
(ii) $E|M_t| < +\infty$;
(iii) For $s < t$, $E_s(M_t) = M_s$. Here E and E_t denote respectively expectation and conditional expectation with respect to $\mathcal{P}_t$. Multidimensional martingales are defined analogously.

A real-valued Brownian motion can also be characterized as a martingale with continuous sample paths such that, for all t, $W_t^2 - t$ is also a martingale.

When a martingale is (a.s.) continuous in time and satisfies the assumption $E|M_t|^2 < +\infty$, we define its quadratic variation as the limit, in probability, of the sums

$$< M, M >_t = \sum_{t_i, t_{i+1}} (M_{t_{i+1}} - M_{t_i})^2$$

when the mesh of the partition $\{t_i\}$ goes to zero. For a Brownian motion defined in $[0, T]$ and starting from zero, this limit is equal to T. One defines the covariation between two martingales M_t and N_t as the limit of the sums

$$< M, N >_t = \sum_{t_i, t_{i+1}} (M_{t_{i+1}} - M_{t_i})(N_{t_{i+1}} - N_{t_i}).$$

A stochastic process X_t is a *semimartingale* if, for every t it can be decomposed into a sum

$$X_t = M_t + A_t$$

of a martingale M_t and a process of bounded variation (or almost-surely differentiable) A_t, with $A_0 = 0$ a.s. The martingale part describes a diffusion, the bounded variation part an evolution which is "similar" to a deterministic one. The quadratic variation of a semimartingale is, by definition, equal to the quadratic variation of its martingale part.

Martingales and, in particular, Brownian motion, are (almost surely) never differentiable in time; therefore one cannot integrate with respect to them with a Lebesgue–Stieltjes kind of integration. Itô introduced a theory of stochastic integration, now named *Itô's stochastic calculus*. The Itô's integral is defined by the following limit

$$\int_0^t X(s) dW(s) = \lim \sum_{t_i, t_{i+1}} X(t_i)(W_{t_{i+1}} - W_{t_i}),$$

the limit being taken in probability and when we consider partitions of the time interval $[0, t]$ with mesh converging to zero. Itô's integral is well defined when X_t is a semimartingale such that $E|X_t|^2 dt < +\infty$.

Notice that the values of the integrand X_t are taken on the left point of the intervals $[t_i, t_{i+1}]$. Unlike in usual Lebesgue–Stieltjes integration, considering its values in any other point of these intervals leads to completely different results. Another common and interesting way to define a stochastic integral is the so-called *Stratonovich integral*:

$$\int X(s) \circ dW(s) = \lim \frac{1}{2} \sum_{t_i, t_{i+1}} (X(t_i) + X_{t_{i+1}})(W_{t_{i+1}} - W_{t_i}).$$

Each type of integrals present its own advantages. Although Stratonovich integral demands more regularity on the integrand in order to be well defined, it is a more intrinsic concept and more adapted to be extended to semimartingales with values in curved spaces, for example. Also, as we shall write below, the rules of differential calculus are analogous to the classical ones for these integrals (and quite different for the Itô integral), which makes Stratonovich somehow popular in applications or in Physics. Nevertheless it turns out that $\int_0^t X(s)dW(s)$ is a martingale, where $\int_0^t X(s)\circ dW(s)$ is not, in general. More important for us, any martingale is in fact a stochastic Itô integral and therefore any semimartingale can be decomposed into an Itô integral representing a diffusion, or a pure fluctuation, and a bounded variation part that represents the mean dynamical content of the process. So even though, when both integrals are defined, they are related by the following formula,

$$\int_0^t X(s)\circ dW(s) = \int_0^t X(s)dW(s) + \frac{1}{2}\int_0^t d < X,W >_s,$$

dynamics is better identified using Itô integration.

Rules of differentiation are given by the so-called *Itô's formula*. If f is a regular function and X_t a ($\mathbb{R}^d$ valued) semimartingale,

$$f(X_t) = f(X_0) + \sum_{i=1}^{d}\int_0^t \frac{\partial}{\partial x_i}(X_s)dX^i(s) + \frac{1}{2}\sum_{i,j=1}^{d}\int_0^t \frac{\partial^2}{\partial x_i\partial x_j}f(X_s)d < X^i, X^j >_s$$

or, in differential form,

$$df(X_t) = \sum_{i=1}^{d}\frac{\partial}{\partial x_i}(X_t)dX^i(t) + \frac{1}{2}\sum_{i,j=1}^{d}\frac{\partial^2}{\partial x_i\partial x_j}f(X_t)d < X^i, X^j >_t .$$

For Stratonovich integrals we have, as in usual differential calculus,

$$df(X_t) = \sum_{i=1}^{d}\frac{\partial}{\partial x_i}(X_t)\circ dX^i(t).$$

If f is time dependent one has to add the term $\frac{\partial f}{\partial t}(t, X_t)dt$ in the formulae.

Recall that, in the case of independent Brownian motions W_t^i, and writing $\mathrm{id}(t) = t$, we have

$$d < W^i, W^j >_t = \delta_{ij}\, dt, \qquad d < W^i, \mathrm{id} >_t = 0 \ \forall i.$$

Stochastic differential equations generalize ordinary differential ones. Given σ with values in matrices of type $d\times r$, a vector field (possibly time-dependent) u and an initial random variable X_0, these equations take the form

$$dX_t = \sigma(X_t).dW(t) + u(t, X_t)dt.$$

Using Itô's formula we deduce that, if X_t is a solution of such an equation and $X_0 = x$, we have, for a smooth function f,

$$\lim_{t\to 0}\frac{1}{t}E_x(f(X_t) - f(0,x)) = \mathcal{L}f(x),$$

where $\mathcal{L}$ is the second order linear operator

$$\mathcal{L}f(x) = \frac{1}{2}\sum_{i,j}(\sigma\sigma^T)_i^j\frac{\partial^2 f}{\partial x_i\partial x_j} + (u.\nabla f).$$

The operator $\mathcal{L}$ is called the generator of the process X_t.

Solutions of stochastic differential equations are particular cases of semimartingales. They are defined globally in time when the coefficients σ and u are smooth and bounded. Many works, even very recent ones, have been devoted to extend existence of solutions for less regular coefficients, in particular of Sobolev type.

When considering diffusion processes on domains rather than in all the Euclidean space, and subject to boundary conditions, we need to use the notion of *local time*. The local time of a Brownian motion is a family of (a.s.) continuous non negative random variables $\ell(t,x)$ such that, for any set A and $t > 0$ we have

$$\int_0^t 1_A(W_s)ds = 2\int_A \ell(t,x)dm(x).$$

It may also be regarded as the limit

$$\ell(t,x) = \lim_{\varepsilon\to 0}\frac{1}{4\varepsilon}\int_0^t 1_{]x-\varepsilon,x+\varepsilon[}(W_s)ds.$$

Local time corresponds to the amount of time spent by a Brownian path in a neighborhood of a point $x \in \mathbb{R}$. This concept was introduced by Paul Lévy in 1948 (see, for example [48]). The concept extends naturally to the multidimensional case as well as to semimartingales X_t. We have,

$$\int_0^t f(X_t)d < X,X >_t = 2\int_{-\infty}^{+\infty} f(x)\ell(t,x)dm(x)$$

for every regular function f, where $\ell(t,x) = \int_0^t 1_{\partial\mathcal{O}}(X_s(x))d\ell(s)$. Then we can consider stochastic differential equations in domains $\mathcal{O} \subset \mathbb{R}^d$, of the form

$$dX_t = \sigma(X_t).dW(t) + u(t,X_t)dt + n(X_t)d\ell(t), \quad X_0 = x,$$

where n denotes the unitary vector normal to the boundary. These stochastic processes are reflected at the boundary and, in terms of partial differential equations (their generators), they correspond to considering Neumann boundary conditions.

Finally we consider the more recent notion of *forward–backward stochastic differential equation* (cf., for example [49]).

For given (smooth) coefficients σ, u, v, initial condition X_0 and final condition h, one looks for semimartingales $X_t, Y_t, t \in [0,T]$ which are solutions of the following system of stochastic differential equations (written here in integral form),

$$\begin{cases} X_t = X_0 + \int_0^t \sigma(X_s).dW_s + \int_0^t u(s,X_s)ds \\ Y_t = h(X_T) - \int_t^T Z_s.dW_s + \int_t^T v(s,X_s,Y_s,Z_s)ds \end{cases}$$

with $E\int_0^T [|X_t|^2 + |Y_t|^2 + |Z_t|^2]dt < \infty$. There are two remarkable features of such systems: one is that, in spite of a condition given at a final time (h), these solutions turn out, in fact, to be adapted to the past filtration. Another one is that, even if a priori we have three unknowns X, Y and Z, the last one will be in fact equal to $Z_t = \nabla u(X_t)$. These kind of systems are natural generalizations of second order ordinary differential equations to the stochastic setting.

References

1. Marchioro, C.; Pulvirenti, M. Vortex methods in two-dimensional fluid mechanics. In *Lecture Notes in Physics*; Springer: Berlin Germany, 1984
2. Vishik, M.I.; Komechi, A.I.; Fursikov, A.I. Some mathematical problems of statistical hydrodynamics. *Russ. Math. Surv.* **1979**, *34*, 149–234. [CrossRef]
3. Gawedzki, K. Soluble models of turbulent transport. In *Non-Equilibrium Statistical Mechanics and Turbulence*; Nazarenko, S., Zaboronski, O., Eds.; Cambridge Uniersity Press: Cambridge, UK, 2008; pp. 47–107.
4. Palmer, T.N.; Williams, P.D. Introduction. Stochastic physics and climate modelling. *Philos. Trans. R. Soc. A* **2008**, *366*, 2421–2427. [CrossRef] [PubMed]
5. Crisan, D.; Flandoli, F.; Holm, D.D. Solution properties of a 3D stochastic Euler Fluid equation. *J. Nonlin. Sci.* **2019**, *29*, 813–870. [CrossRef]

6. Bensoussan, A.; Teman, R. Équations stochastiques du type Navier–Stokes. *J. Funct. Anal.* **1973**, *13*, 195–222. [CrossRef]
7. Pope, S.B. On the relationship between stochastic Lagrangian models of turbulence and second-moment closures. *Phys. Fluids* **1994**, *6*, 973–985. [CrossRef]
8. Holm, D.D. Variational principles for stochastic fluid dynamics. *Proc. R. Soc. A* **2015**, *471*, 20140963. [CrossRef]
9. Busnello, B. A probabilistic approach to the two-dimensional Navier–Stokes equation. *Ann. Probab.* **1999**, *27*, 1750–1780. [CrossRef]
10. le Jan, Y.; Sznitman, A.S. Stochastic cascades and the 3-dimensional Navier–Stokes equations. *Probab. Theory Relat. Fields* **1997**, *109*, 343–366. [CrossRef]
11. Constantin, P.; Iyer, G. A stochastic Lagrangian representation of the three-dimensional incompressible Navier–Stokes equation. *Commun. Pure Appl. Math.* **2008**, *61*, 330–345. [CrossRef]
12. Nakagomi, T.; Yasue, K.; Zambrini, J.-C. Stochastic variational derivations of the Navier–Stokes equation. *Lett. Math. Phys.* **1981**, *160*, 337–365. [CrossRef]
13. Yasue, K. A variational principle for the Navier–Stokes equation. *J. Funct. Anal.* **1983**, *51*, 133–141. [CrossRef]
14. Arnold, V.I. Sur la géométrie différentielle des groupes de Lie de dimension infinie et ses applications à l'hydrodynamique des fluides parfaits. *Ann. Inst. Fourier* **1966**, *16*, 316–361. [CrossRef]
15. Ebin, D.G.; Marsden, J.E. Groups of diffeomorphisms and the motion of an incompressible fluid. *Ann. Math.* **1970**, *17*, 102–163. [CrossRef]
16. Arnold, V.I.; Khesin, B.A. *Topological Methods in Hydrodynamics*; Springer: Berlin, Germany, 1998.
17. Eyink, K.L. Stochastic line motion and stochastic flux conservation for nonideal hydromagnetic models. *J. Math. Phys.* **2009**, *50*, 083102. [CrossRef]
18. Koide, T.; Kodama, T. Navier–Stokes, Gross-Pitaevskii and generalized diffusion equations using the stochastic variational method. *J. Math. A* **2012**, *45*, 255204. [CrossRef]
19. Ikeda, N.; Watanabe, S. *Stochastic Differential Equations and Diffusion Processes*; Springer: Berlin, Germany, 1981.
20. Marsden, J.E.; Ratiu, T.S. Introduction to Mechanics and Symmetry: A Basic Exposition of Classical Mechanical Systems. In *Texts in Applied Mathematics*; Springer: Berlin, Germany, 1999; Volume 17.
21. Nelson, E. *Dynamical Theories of Brownian Motion*; Princeton University Press: Princeton, NJ, USA, 1967.
22. Cipriano, F.; Cruzeiro, A.B. Navier–Stokes equation and diffusions on the group of homeomorphisms of the torus. *Commun. Math. Phys.* **2007**, *275*, 255–269. [CrossRef]
23. Arnaudon, M.; Cruzeiro, A.B. Lagrangian Navier–Stokes diffusions on manifolds: variational principle and stability. *Bull. Sci. Math.* **2012**, *136*, 857–881. [CrossRef]
24. Arnaudon, M.; Chen, X.; Cruzeiro, A.B. Stochastic Euler-Poincaré reduction. *J. Math. Phys.* **2014**, *55*, 081507. [CrossRef]
25. Chen, X.; Cruzeiro, A.B.; Ratiu, T.S. Stochastic variational principles for dissipative equations with advected quantities. *arXiv* **2018**, arXiv:1506.05024.
26. Cruzeiro, A.B. Navier–Stokes and stochastic Navier–Stokes equations via Lagrange multipliers. *J. Geom. Mech.* **2019**, *11*, 553–560. [CrossRef]
27. Latas, M. On the Derivation of the Navier–Stokes Equations from a Nondeterministic Variational Principle. Master's Thesis, University of Lisbon, Lisbon, Portugal, 2019.
28. Cruzeiro, A.B.; Shamarova, E. Navier–Stokes equations and forward-backward SDEs on the group of diffeomorphisms of a torus. *Stoch. Proc. Their Appl.* **2009**, *119*, 4034–4060. [CrossRef]
29. Cruzeiro, A.B.; Qian, Z. Backward stochastic diferential equations associated with the vorticity equations. *J. Funct. Anal.* **2014**, *267*, 660–677. [CrossRef]
30. Chen, X.; Cruzeiro, A.B.; Qian, Z. Navier–Stokes equation and forward-backward stochastic differential system in the Besov spaces. *arXiv* **2013**, arXiv:1305.0647.
31. Douglas, J.; Ma, J.; Protter, P. Numerical methods for forward-backward stochastic differential equations. *Ann. Appl. Prob.* **1996**, *6*, 940–968. [CrossRef]
32. Lejay, A.; González, H.M. A Forward-Backward Probabilistic Algorithm for the Incompressible Navier–Stokes Equations. 2019. Available online: https://hal.inria.fr/hal-02377108 (accessed on 26 February 2020).
33. Chen, X.; Cruzeiro, A.B. Stochastic geodesics and forward-backward stochastic differential equations on Lie groups. *Discrete Contin. Dyn. Syst.* **2013**, 115–121. [CrossRef]

34. Arnaudon, M.; Cruzeiro, A.B.; Léonard, C.; Zambrini, J.-C. An entropic interpolation problem for incompressible viscid fluids. *Ann. Inst. H. Poincaré*, **2019**.
35. Schrödinger, E. Sur la théorie relativiste de l'electron et l'interprétation de la mécanique quantique. *Ann. Inst. H. Poincaré* **1932**, *2*, 269–310.
36. Zambrini, J.-C. Stochastic mechanics according to E. Scrödinger. *Phys. Rev. A* **1986**, *33*, 1532–1548. [CrossRef]
37. Léonard, C. A survey of the Scrödinger problem and some of its connections with optimal transport. *Discrete Contin. Dyn. Syst.* **2014**, *34*, 1533–1574. [CrossRef]
38. Villani, V. *Optimal Transport. Old and New*; Springer: Wissen, Germany, 2009.
39. Brenier, Y. A homogenized model for vortex sheets. *Arch. Ration. Mech. Anal.* **1997**, *138*, 319–353. [CrossRef]
40. Brenier, Y. The least action principle and the related concept of generalized flows for incompressible perfect fluids. *J. Am. Math. Soc.* **1989**, *2*, 225–255. [CrossRef]
41. Arnaudon, M.; Cruzeiro, A.B.; Fang, S. Generalized stochastic Lagrangian paths for the Navier–Stokes equation. *Ann. Sci. Norm. Super. Pisa Cl. Sci.* **2018**, *18*, 1033–1060.
42. Arnaudon, M.; Cruzeiro, A.B.; Galamba, N. Lagrangian Navier–Stokes flows: a stochastic model. *J. Phys. A Math. Theory* **2011**, *44*, 175501. [CrossRef]
43. Cruzeiro, A.B.; Liu, G. A stochastic variational approach to the viscous Camassa-Holm and Leray-alpha equations. *Stoch. Proc. Their Appl.* **2017**, *127*, 1–19. [CrossRef]
44. Holm, D.D.; Marsden, J.E.; Ratiu, T.S. The Euler-Poincaré equations and semidirect products with applications to continuum mechanics. *Adv. Math.* **1998**, *137*, 1–81. [CrossRef]
45. Khesin, B.; Misiolek, G.; Modin, K. Geometric hydrodynamics via Mandelung transform. *PNAS* **2018**, *115*, 6165–6170. [CrossRef]
46. Thieullen, M.; Zambrini, J.-C. Probability and quantum symmetries. I. The theorem of Noether in Schrödinger's Euclidean quantum mechanics. *Ann. Inst. H. Poincaré Phys. Théorique* **1997**, *67*, 297–338.
47. Cruzeiro, A.B.; Lassalle, R. Symmetries and martingales in a stochastic model for the Navier–Stokes equation. In *From Particle Systems to PDEs III*; Springer: Berlin, Germany, 2016; Volume 162, pp. 185–194.
48. Pilipenko, A. *An Introduction to Stochastic Differential Equations with Reflection*; Postdam University Press: Postdam, Germany, 2014.
49. Ma, J.; Yong, J. Forward-backward stochastic differential equations and their applications. In *Lecture Notes in Math*; Springer: Berlin, Germany, 2007.

Technical Note

Damage Characteristics and Mechanism of a 2010 Disastrous Groundwater Inrush Occurred at the Luotuoshan Coalmine in Wuhai, Inner Mongolia, China

Fangpeng Cui [1,2,*], Qiang Wu [1,2], Chen Xiong [1,2,*], Xiang Chen [1,2], Fanlan Meng [1,2] and Jianquan Peng [1,2]

1 College of Geoscience and Surveying Engineering, China University of Mining and Technology, Beijing 100083, China; wuq@cumtb.edu.cn (Q.W.); chenxiang@student.cumtb.edu.cn (X.C.); meng@student.cumtb.edu.cn (F.M.); fitz_peng@student.cumtb.edu.cn (J.P.)

2 National Engineering Research Center of Coal Mine Water Hazard Controlling, Beijing 100083, China

* Correspondence: cuifp@cumtb.edu.cn (F.C.); xiongc@student.cumtb.edu.cn (C.X.)

Received: 12 February 2020; Accepted: 25 February 2020; Published: 28 February 2020

Abstract: On 1 March 2010, a disastrous groundwater inrush occurred at the Luotuoshan coalmine in Wuhai (Inner Mongolia, China). Great effort was taken during the post-accident rescue. However, triggered by a large amount of groundwater rushed in from the Ordovician limestone aquifer underlying the No.16 coal seam through the fractured sandy claystone and the karst collapse column, it caused great damage, including 32 deaths and direct economic losses of over 48 million yuan. The groundwater inrush originated from the floor heave in the air return gallery of the No.16 coal seam. The peak inflow rate was 60,036 m^3/h. The gallery excavation under conditions caused by the incompletely recognized hydrogeological environment induced the accident. The unidentified spatial distribution of the karst collapse column triggered the accident directly. The high-pressure groundwater accumulated in the collapse column and the gallery excavation, which caused the redistribution of the in situ stress, contributing to progressive fractures in the floor of the No. 16 coal seam. Eventually, an intensive water-conductive passage consisting of the fractured floor and the karst collapse column formed. Administratively/technically, that mandatory regulations on gallery excavation were not carried out which contributed the accident. Moreover, the poor awareness about groundwater inrush recognition and quick remediation also contoirbuted to the disastrous extent of the accident.

Keywords: groundwater inrush; the Luotuoshan coalmine; damage mechanism; karst collapse column

---

1. Introduction

Underground gallery excavation and coal mining can severely influence the nearby strata. Meanwhile, the nearby hydrogeological conditions will also change seriously. As a result, groundwater inrushes into mining areas occur sometimes all over the world. Cases happening in the United Kingdom, Italy, Slovenia, Poland, Australia, India, etc., have been reported and the causes discussed [1–5]. Eventually, for mitigating their great damage, many effective methods were proposed [6–8].

In China, to meet great requirement of the rapid economic progress in recent decades, annual coal production has increased and remained at a very high level. Coalmine water inrushes have occurred sometimes during this process. Figure 1 details some properties of the groundwater inrushes from 2000. The number of the accidents and deaths obviously decreased, but unfortunately, some serious groundwater inrush episodes brought great damage [9,10]. Therefore, there is still a long way to go for their prevention.

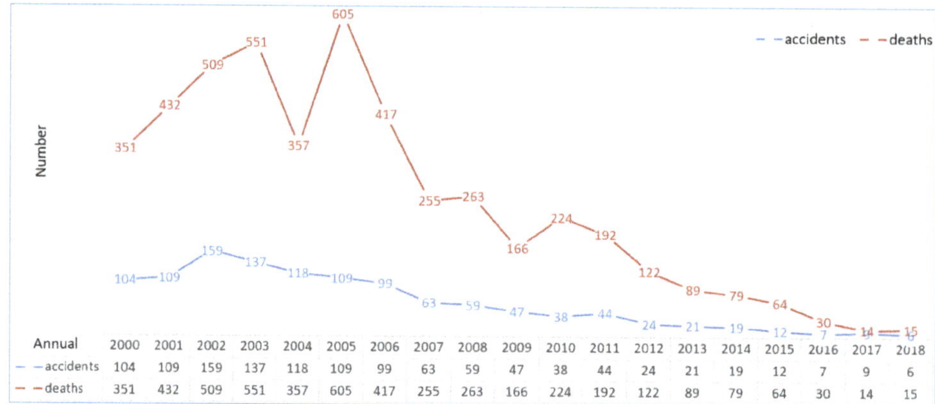

Figure 1. Characteristics of coalmine water disasters occurred in China from 2000 to 2018.

Actually, in China, a great number of related studies have been conducted. Causes, geological circumstances, categories, early recognition, forecasting, prevention and control methods for the inrushes were the focus in these studies [11–16]. Subsequently, some research-based and practice-tested techniques and methods formed gradually [17–21]. All of these effectively strengthen safety of underground coal mining operations.

Nevertheless, at 7:20 am on 1 March 2010, a disastrous groundwater inrush occurred at the Luotuoshan coalmine located in Wuhai, Inner Mongolia, China. When the accident happened, the coalmine was in gallery-construction mode and 77 miners were working in galleries. Fortunately, 45 miners escaped from the accident, but 32 miners were drowned. Furthermore, the accident caused direct economic losses of over 48 million Yuan.

2. The Coalmine

Located in Wuhai (Figure 2), Inner Mongolia, the Luotuoshan coalmine began its construction in 2006. Its annual design capacity was 1.5 million tons of raw coal. The coalfield is about 10-km long along the north-south direction and 4-km wide along the east-west direction. Its area is about 38.7 km^2. The coal reserves are estimated at about 0.42 billion tons and the exploitable amount is about 0.25 billion tons.

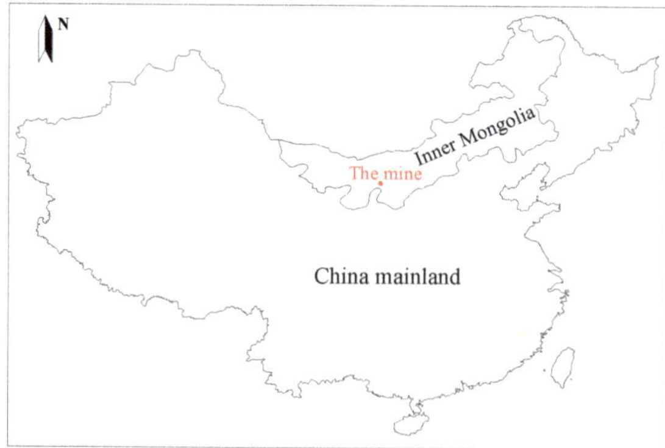

Figure 2. Location sketch of the Luotuoshan coal mine.

Before the date of the groundwater inrush, construction of the main inclined shaft, auxiliary inclined shaft and vertical air return shaft of the coalmine had been completed. Meanwhile, construction of main galleries in the No. 9 zone and the No. 16 coal seam had been completed, too. As shown in Figure 3, at the time of the groundwater inrush, two air return galleries in the No. 9 coal seam and an air return gallery in No. 16 coal seam were under construction.

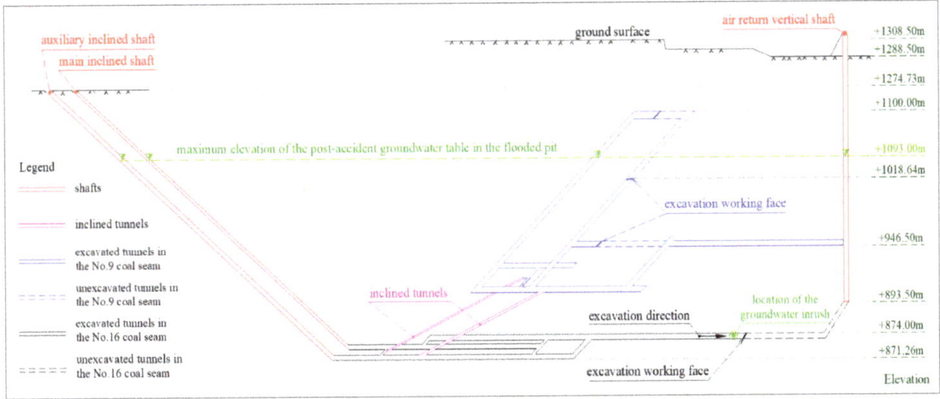

Figure 3. Distribution sketch of the shafts and galleries in the Luotuoshan pit.

3. The Geological Setting

The Luotuoshan coalfield is located in a semi-desert region. Mountains and valleys exist on its east and west boundary. It rains a little and a drought climate prevails. Surface runoff forms scarcely. The Yellow River passes through the west of the coalfield. Normal faults develop in the shallow ground of the coalfield.

Sand beds and gravels cover the surface. As shown in Figure 4, the formed strata in the coalfield, in chronological order, include: (1) karst limestone formed in the Ordovician Period, (2) coal seam, sandy claystone, sandstone and claystone formed in the Carboniferous Period, (3) coal seam, sandy claystone, sandstone and claystone formed in the Permian Period, (4) sandy claystone and coarse sandstone formed in the Triassic Period and, (5) sand bed and gravel formed in the Quaternary Period. Thickness of the Ordovician strata is about 200 m. Thickness of the Carboniferous strata is about 40–100 m. Total thickness of the Permian and Triassic strata is about 236–897 m. Thickness of the Quaternary strata is about 0–25 m.

Coal seams in the Luotuoshan coalfield are located mainly in the Permian and the Carboniferous strata. The No. 9, the No. 10 and the No. 16 coal seam are the main minable ones. Their total thickness is about 10.41 m. Their properties are detailed as follows: the thickness of the No. 9 coal seam varies from 0.94 m to 8.24 m and the average is about 4.32 m. Its roof and floor are comprised of claystone. The average thickness of the No. 10 coal seam is about 0.98 m. The average thickness of the No. 16 coal seam is about 5.13 m. Its roof is comprised of sandy claystone, claystone and sandstone. Its floor is comprised of siltstone, fine sandstone, sandy claystone and claystone.

Based on lithological properties, three types of aquifers developed in the coalfield. They are the Quaternary porous phreatic aquifer, Permian, Carboniferous fractured sandstone confined aquifer, and Ordovician karst limestone confined aquifer.

Alluvium, diluvium, aeolian sand, loessial sand and colluvium compose the Quaternary phreatic aquifer. Its maximum total thickness is about 25 m and the average is about 6 m. The alluvium and diluvium are located at the dry river valleys and terraces with different elevations. They comprise sand and gravel layers. Pores are partly filled with groundwater. The depth of the groundwater varies from 1.15 to 19.13 m. Because of their limited distribution in space, aeolian sand, loessial sand and colluvium contain litter groundwater. They are recharged directly by rainfall. Meanwhile, they recharge the neighbor underlying strata.

strata system	average thickness / m	depth / m	stratigraphic column	lithology	aquifers
Quaternary	5.97	5.97		sand bed & gravel	porous phreatic aquifer
Triassic	73.50	79.47		sandy claystone	
	97.20	176.67		coarse sandstone	
	16.07	192.74		sandy claystone	
	25.05	217.79		coarse sandstone	
Permian	50.61	268.40		sandy claystone sandstone No.1 coal seam	fractured confined aquifer
	8.32	276.72		sandy claystone No.2 & 3 coal seam	
	30.08	306.80		coarse sandstone No.8 coal seam	fractured confined aquifer
	14.45	321.25		claystone No.9 & 10 coal seam	
Carboniferous	34.19	355.44		sandy claystone claystone No.12 coal seam	fractured confined aquifer
	34.18	389.62		sandy claystone sandstone No.14 15 16 & 17 coal seam	location of the water inrush
	18.11	407.73		sandstone No.18 & 19 coal seam	fractured confined aquifer
Ordovician	>200.0			karst limestone	karst confined aquifer

Figure 4. Strata and aquifer distribution at the Luotuoshan coal mine.

The fractured sandstone-confined aquifers comprise four parts. Their properties are detailed as follows. As shown in Figure 4, the first is located in Late and Middle Permian strata overlying the No. 1 coal seam. Its thickness varies from 144 m to 487 m. The elevation of the groundwater table is about +1262.25 m. The second is located in Early Permian strata between the No. 3 and the No. 8 coal seams. Its thickness is about 15 m. The elevation of the groundwater table varies from +1295.93 m to +1272.12 m. The third is located in the Carboniferous strata between the No. 10 and the No. 16 coal seams. Its thickness is about 11.5 m. The elevation of the groundwater table varies from +1210.67 m to +1274.02 m. The fourth is located in the Carboniferous strata between the No. 16 and the No. 18 coal seams. Its thickness is about 15 m. The elevation of the groundwater table is about +1276.50 m.

Finally, the karst-confined aquifer is located in the Ordovician limestone. Its thickness is bigger than 200 m. Elevation of the groundwater table varies from +1117.75 m to +1070 m. The water-abundance is proportional to the porosity. It possesses a high water-abundance in the porous area and a low water-abundance in the intact area.

The average vertical distance between the No. 16 coal seam and top of the Ordovician limestone is only about 20 m. As a result, the top of the karst aquifer will probably heave because of the goaf formed during the underground coal mining. Furthermore, groundwater in the karst aquifer will probably rush into the goaf when any fractures induced by the heave meet the goaf. The probable groundwater inrush will seriously damage the underground mining.

4. The Accident Sequence and the Rescue Operations

At 5:50, 1 March 2010, a jet stream gushed out from a blast hole in the heading face of the air return gallery in the No. 16 coal seam. It continued for approximate 5 s.

At 6:10, more jet streams appeared from 5-6 blast holes. The maximum diameter of the water streams was approximate 20 cm. Meanwhile, a small part of the floor heaved 0.1-0.3 m. This part was 10-m long and 1.2-m wide. As shown in Figure 3, it originated from a location where the distance between it and the heading face was approximate 26 m. An obvious fracture appeared at the boundary of the heaved part. The fracture was approximate 10-m long, 4-cm wide and 5-cm deep.

At 6:30, more and more blast holes were filled with gushing streams. The height of the heaved part was increasing.

At 7:20, a great amount of groundwater gushed out and galleries in the No. 16 coal seam were flooded. The depth of the accumulated water reached approximately 1 m. At that moment, miner evacuation and rescue operations were initiated immediately.

At 8:05, the accumulated groundwater reached bottom of the vertical air return shaft with an elevation of +893.50 m.

At 14:00, the elevation of the accumulated groundwater reached +1079.05 m. Galleries located in No. 16 and No. 9 coal seams were almost submerged.

At 24:00, 3 March, the maximum elevation of the accumulated groundwater table in the pit reached +1093.00 m, as Figure 3 illustrates.

During the post-accident rescue, large-scale drainage was performed. Meanwhile, surface boreholes were conducted to connect the galleries for probable survivors. At 09:24, 7 March, six boreholes connected the galleries in the No. 9 coal seam with an elevation of +1100 m. Unfortunately, no survivors was found. After that, the drilling ended. Grouting and continuous drainage were conducted for the coalmine recovery. The grouting ended on 28 April. The drainage ended on 10 May. Underground rescue started on 14 April and ended on 7 May. Finally, 32 drowned miners were found in the pit.

5. Characteristics and Mechanism of the Groundwater Inrush

5.1. The Source

According to recall of some of the miners who escaped from the flooded pit and post-accident site-reconnaissance, the groundwater inrush originated from the floor heave, in the air return gallery in the No. 16 coal seam, as Figure 5 illustrates.

The inburst groundwater is from the Ordovician limestone aquifer, which underlies the No. 16 coal seam. The reasons are as follows: Firstly, according to the hydrochemical test results, the properties of the inburst groundwater are similar to those of the nearby Ordovician karst groundwater, as shown in Table 1. Secondly, based on monitoring records of tables of the accumulated groundwater in the pit and drainage flow rates in the post-accident rescue, as Figure 6 illustrates, variations of the two curves show that the inburst groundwater behaved with obvious characteristics of the Ordovician karst groundwater. This conclusion is supported by the following facts: ① During the initial drainage period, i.e., the *a-b* segment in Figure 6, the groundwater table still increased slowly even if the drainage flow rate remained a certain constant scale. This showed that the groundwater in the pit was recharged strongly and continuously. The recharge behaved with obvious characteristics of the Ordovician karst groundwater. ② The groundwater table in the pit decreased obviously on increasing of the drainage flow rate. Then the decrease developed slowly because of the increased recharge, i.e., *b-c* and *f-g* segments in Figure 6. It behaved with obvious characteristics of the Ordovician karst groundwater, too. ③ The groundwater table recovered obviously on decreasing of the drainage flow rate, i.e., *c-d*, *e-f* and *h-i* segments in Figure 6. It showed that the groundwater in the pit was recharged strongly and continuously. The recharge behaved with obvious characteristics of the Ordovician karst groundwater, too.

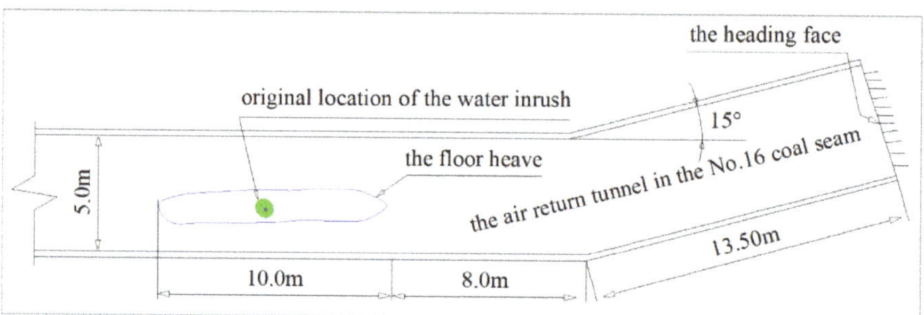

Figure 5. A plane sketch of the origin of the groundwater inrush.

Table 1. Hydrochemical properties of the Ordovician karst groundwater from different coalmines/places.

Sampling Site	Drained-out Groundwater during the Post-Accident Rescue	The Ordovician Karst Groundwater Sampled in the Galleries of the No.16 Coal Seam before the Accident	The Ordovician Karst Groundwater Sampled at a Nearby Coalmine	The Ordovician Karst Groundwater Sampled in a nearby Supply Well
Sampling date	4 March 2010	1 January 2008	6 March 2010	7 March 2010
Total hardness	284.50	266.24	399.3	494.60
Carbonate hardness	194.88	199.18	220.2	179.64
Ca^{2+}	65.64/3.27	57.11/2.85	93.2/4.65	116.35/5.81
Mg^{2+}	29.29/2.41	30.02/2.47	40.5/3.33	49.97/4.11
Na^{+}	85.45/3.72	73.16/3.18	91.2/3.97	84.02/3.65
K^{+}	4.77/0.12	3.18/0.08	5.50/0.14	3.50/0.09
Total cation	185.25/9.53	163.47/8.58	230.4/12.09	253.81/13.66
Cl^{-}	95.42/2.69	80.83/2.28	108.1/3.05	94.06/2.67
SO_4^{-}	141.26/2.94	87.42/1.82	194.1/4.14	350.52/7.30
HCO_3^{-}	237.61/3.89	242.85/3.98	268.5/4.40	219.01/3.59
F^{-}	0.65/0.03	0.43/0.02	0.50	1.02/0.05
Total anion	474.41/9.53	427.32/8.35	580.70/11.65	664.36/13.65
PH value	6.9	7.5	7.5	7.82
Total solid	700	600.12	684.95	429.87
Sample temperature	20.5 °C	25 °C	19 °C	20 °C

Unit: $mg \cdot L^{-1}/mmol \cdot L^{-1}$.

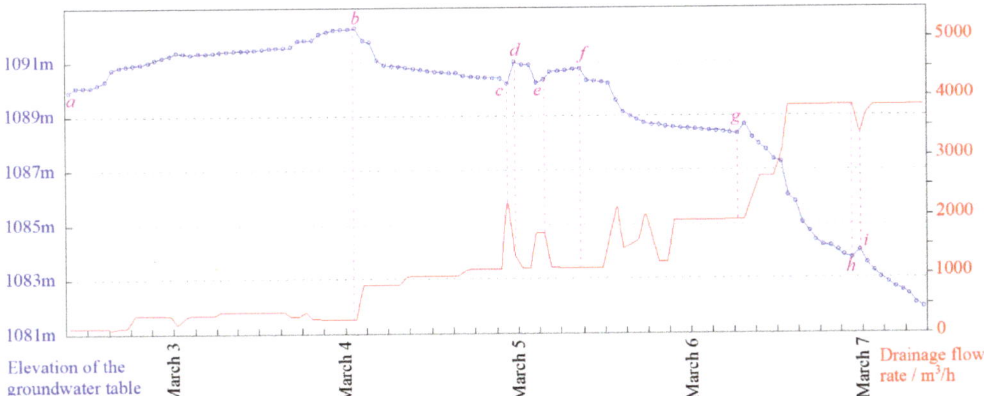

Figure 6. Monitoring records of the groundwater table and the drainage flow rate in the post-accident rescue.

Furthermore, according to volume of the flooded galleries and shafts, i.e., approximate 67,000 m^3, from 7:30 to 8:40 on 1 March, an estimated peak flow rate of the groundwater inrush was 60,036 m^3/h. It behaved an obvious characteristic of the Ordovician karst groundwater inrush.

5.2. The Passage and the Water Inrush Mechanism

With respect to the passage, as shown in Figure 7, a water-conductive karst collapse column was uncovered by some boreholes drilled after the site rescue. The collapse column, the gallery excavation and the high-pressure groundwater induced together the passage to form gradually.

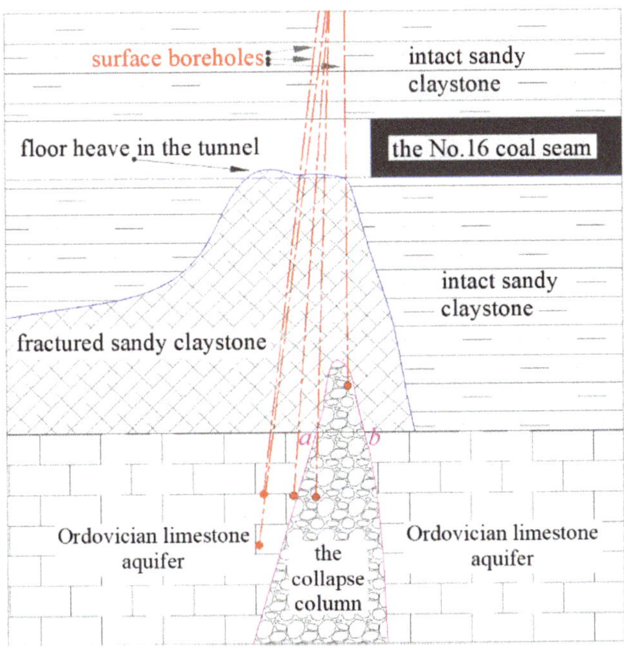

Figure 7. Distribution sketch of the passage of the inburst groundwater.

Firstly, the vertical distance between the floor of the No. 16 coal seam and the top of the Ordovician limestone was approximate 34 m. However, according to the site tests and theoretical calculations, the safe thickness of the aquiclude was approximate 11.5 m. Therefore, the karst groundwater had no way to touch the floor of the No.16 coal seam under tectonic influence conditions or absent a geological anomaly. Secondly, based on the borehole surveying and the related analysis, as Figure 7 illustrates, excavation of the air-return gallery of the No.16 coal seam triggered the underlain sandy claystone to fracture. Therefore, the floor heave formed gradually. Furthermore, the high-pressure groundwater accumulated in the water-conductive collapse column induced the overlying sandy claystone to fracture, too. The fractured sandy claystone expanded gradually and eventually the water-conductive passage formed. As a result, the high-pressure groundwater in the Ordovician limestone aquifer rushed into the galleries in a large scale and the disastrous accident occured quickly.

According to the drilling surveys, the width of the top of the collapse column, *a-b* segment in Figure 7, was approximately 10 m. As shown in Figure 8, core samples in the collapse column are seriously fractured, poorly sorted and poorly rounded. They present obvious properties of the collapse mass.

Figure 8. Core samples obtained from the collapse column.

6. Causes of the Accident

First of all, the gallery excavation contrary to related regulations triggered the disastrous accident. The excavation was conducted under conditions where the hydrogeological environment was not fully recognized. No detailed distribution of the collapse column underlying the No. 16 coal seam was detected and confirmed during the excavation. The conduct was against mandatory items in *Regulations on Coal Mine Water Hazard Controlling* issued by National Coal Mine Safety Administration in 2009. According to the regulations, comprehensive exploration must be carried out before a gallery's construction. Furthermore, effective countermeasures such as grouting, drainage, reservation of waterproof coal (rock) pillar, can be used to control a potential groundwater inrushes triggered by the appearance of a geological anomaly or complications due to hydrogeological conditions. However, in this case the necessary exploration to secure the gallery excavation was not conducted. Therefore, high-pressure groundwater accumulated in the collapse column and the gallery excavation, which caused the redistribution of the in situ stress that contributed to progressive fracturing in the floor of the No. 16 coal seam. Eventually, an intensive passage formed consisting of the fractures and the collapse column. The high-pressure groundwater from the underlying Ordovician limestone aquifer rushed into the galleries immediately and strongly.

Second, some other technical mistakes also contributed the disastrous accident. They are detailed as follows: ① an immediate response was absent. From 5:50 to 7:20 on March 1, i.e., from the time obvious signals appeared to that of a great amount of groundwater inrushing, an evacuation order was not given quickly. The delayed retreat was against the related regulations. Furthermore, it caused great disastrous damage. ② Precision of the hydrogeological achievements didn't secure the gallery excavation and underground mining in the future completely. Actually, some geological explorations were conducted and finished during other galleries' excavation. However, no geological anomaly, such as a collapse column, was found. Most of all, during the excavation, water-abundance exploration and potential release tests were not carried out although they were the most effective measures to secure the excavation.

7. Conclusions and Suggestions

On 1 March 2010, a disastrous groundwater inrush occurred at the Luotuoshan coal mine in Wuhai, Inner Mongolia, China. The accident caused great damage, including 32 deaths and direct economic losses of over 48 million yuan.

The groundwater inrush originated from the floor heave in the air return gallery in the No. 16 coal seam. Results of the hydrochemical tests and relevant records showed that the inburst groundwater was from the Ordovician limestone aquifer underlying the No. 16 coal seam. An estimated peak flow rate of the groundwater inrush was 60,036 m^3/h. It behaved with obvious characteristics of an Ordovician karst groundwater inrush.

Results of subsequent drilling explorations and related theoretical analysis showed that the passage consisted of the fractures formed in the sandy claystone underlying the No.16 coal seam and the collapse column developed in the Ordovician limestone. The high-pressure groundwater accumulated in the collapse column and the gallery excavation, which caused a redistribution of the in situ stress, contributing to progressive fracturing in the floor of the No. 16 coal seam. Eventually, an intensive water-conductive passage formed. The groundwater from the underlain Ordovician limestone aquifer rushed into the galleries immediately and strongly.

The gallery excavation without necessary site exploration was against relevant regulations and triggered the disastrous accident. Furthermore, the poor awareness of groundwater inrush recognition and responses also contributed to the disastrous extent of the accident.

Some important lessons can be learned from this event, as follows: Some important regulations, for example, *Regulations on Coal Mine Water Hazard Controlling*, must be obeyed thoroughly during underground coal mining. The awareness about groundwater inrush recognition and control must be strengthened. Some treatments must be conducted immediately on the site.

Author Contributions: Conceptualization, F.C. and Q.W.; methodology, Q.W.; validation, F.C.; formal analysis, X.C. and F.M.; investigation, X.C., F.M. and J.P.; writing—original draft preparation, F.C. and C.X.; writing—review and editing, Q.W.; visualization, F.C., C.X., F.M. and J.P.; supervision, Q.W.; project administration, Q.W.; funding acquisition, F.C. All authors have read and agreed to the published version of the manuscript.

Funding: This research was funded by the National Key Research and Development Programs of China, grant number 2017YFC0804104 & 2018YFC1504802 and Fundamental Research Funds for the Central Universities of China, grant number 2009QD14.

Acknowledgments: We gratefully acknowledge the financial support from the National Key Research and Development Programs of China (Grant nos. 2017YFC0804104 & 2018YFC1504802), Fundamental Research Funds for the Central Universities of China (2009QD14). We indeed appreciate the experts from the accident investigation team for their great help.

Conflicts of Interest: The authors declare no conflict of interest.

References

1. Bukowski, P. Evaluation of water hazard in hard coal mines in changing conditions of functioning of mining industry in Upper Silesian coal basin-USCB (Poland). *Arch. Min. Sci.* **2015**, *60*, 455–475. [CrossRef]
2. Dash, A.K.; Bhattacharjee, R.M.; Paul, P.S. Lessons learnt from Indian inundation disasters: An analysis of case studies. *Int. J. Disaster Risk Reduct.* **2016**, *20*, 93–102. [CrossRef]
3. Kuscer, D. Hydrological regime of the water inrush into the Kotredez coal mine (Slovenia, Yugoslavia). *Mine Water Environ.* **1991**, *10*, 93–102.
4. Polak, K.; Ro´zkowski, K.; Czaja, P. Causes and effects of uncontrolled water inrush into a decommissioned mine shaft. *Mine Water Environ.* **2016**, *35*, 128–135. [CrossRef]
5. Sammarco, O. Spontaneous inrushes of water in underground mines. *Int. J. Mine Water* **1986**, *5*, 29–42. [CrossRef]

6. Dumpleton, S.; Robins, N.S.; Walker, J.A.; Merrin, P.D. Mine water rebound in South Nottinghamshire: Risk evaluation using 3-D visualization and predictive modelling. *Q. J. Eng. Geol. Hydrogeol.* **2001**, *34*, 307–319. [CrossRef]
7. Kim, J.M.; Parizek, R.P.; Elsworth, D. Evaluation of fully-coupled strata deformation and groundwater flow in response to longwall mining. *Int. J. Rock Mech. Min. Sci.* **1997**, *34*, 1187–1199. [CrossRef]
8. Sammarco, O. Inrush prevention in an underground mine. *Int. J. Mine Water* **1988**, *7*, 43–52. [CrossRef]
9. Cui, F.P.; Wu, Q.; Zhang, S.; Wu, N.; Ji, Y. Damage characteristics and mechanism of a strong water inrush disaster at the wangjialing coal mine, Shanxi province, China. *Geofluids* **2018**, *2018*, 11. [CrossRef]
10. Cui, F.P.; Wu, Q.; Lin, Y.H.; Zeng, Y.F.; Zhang, K.L. Damage features and formation mechanism of the strong water inrush disaster at the daxing co mine, Guangdong province, China. *Mine Water Environ.* **2018**, *37*, 346–350. [CrossRef]
11. Li, S.; Liu, B.; Nie, L.; Liu, Z.; Tian, M.; Wang, S.; Su, M.; Guo, Q. Detecting and monitoring of water inrush in tunnels and coal mines using direct current resistivity method: A review. *J. Rock Mech. Geotech. Eng.* **2015**, *7*, 469–478. [CrossRef]
12. Qian, M.G.; Miao, X.X.; Xu, J.L. Theoretical study of key stratum in ground control. *J. China Coal Soc.* **1996**, *1*, 225–230.
13. Shi, L.; Singh, R.N. Study of mine water inrush from floor strata through faults. *Mine Water Environ.* **2001**, *20*, 140–147. [CrossRef]
14. Wu, Q.; Cui, F.P.; Zhao, S.Q.; Liu, S.Q.; Zen, Y.F.; Gu, Y.W. Type classification and main characteristics of mine water disasters. *J. China Coal Soc.* **2013**, *38*, 561–565.
15. Wu, Q.; Liu, Y.; Zhou, W.; Li, B.; Zhao, B.; Liu, S.; Sun, W.; Zeng, Y. Evaluation of water inrush vulnerability from aquifers overlying coal seams in the Menkeqing coal mine, China. *Mine Water Environ.* **2015**, *34*, 258–269. [CrossRef]
16. Zhang, J.C.; Liu, T.Q. On depth of fissured zone in seam floor resulted from coal extraction and its distribution characteristics. *J. China Coal Soc.* **1990**, *15*, 225–230.
17. Cui, F.P.; Wu, Q.; Lin, Y.H.; Zhao, S.Q.; Zeng, Y.F. Prevention and control techniques & methods for water disasters at coal mines in China. *J. Min. Sci. Technol.* **2018**, *3*, 219–228.
18. Guo, W.J.; Liu, Y.X. Concept and application of floor water inrush coefficient. *Hebei Coal* **1989**, *2*, 56–60.
19. Wang, Z.M.; Lai, Y.W.; Duan, J.J. Application of five figure double coefficient method in water burst evaluation of beixinyao mine floor. *Coal Chem. Ind.* **2016**, *39*, 7–12.
20. Wu, Q.; Hao, X.L.; Dong, D.L. Three maps-two predictions" method to evaluate water bursting conditions on roof coal. *J. China Coal Soc.* **2000**, *25*, 60–65.
21. Wu, Q.; Zhang, Z.L.; Zhang, S.Y.; Ma, J.F. A new practical methodology of the coal floor water bursting evaluating II: The vulnerable index method. *J. China Coal Soc.* **2007**, *32*, 1121–1126.

MDPI
St. Alban-Anlage 66
4052 Basel
Switzerland
Tel. +41 61 683 77 34
Fax +41 61 302 89 18
www.mdpi.com

Water Editorial Office
E-mail: water@mdpi.com
www.mdpi.com/journal/water

www.ingramcontent.com/pod-product-compliance
Lightning Source LLC
LaVergne TN
LVHW070649170726
843515LV00004B/363

* 9 7 8 3 0 3 9 4 3 7 4 7 4 *